D0364558

FLIGHTS TO DISASTER

FLIGHTS TO DISASTER

ANDREW BROOKES

Contents

First published 1996

ISBN 0 7110 2475 8

Published by Ian Allan Publishing

an imprint of Ian Allan Ltd, Terminal House, Station Approach, Shepperton, Surrey TW17 8AS. Printed by Ian Allan Printing Ltd, Coombelands House, Coombelands Lane, Addlestone, Surrey KT15 1HY.

For
Charlie and Kathy

Front cover:
On 4 October 1992, a Boeing 747 freighter belonging to El Al Israel Airlines took-off from Amsterdam's Schipol airport bound for Tel Aviv. As the cargo aircraft climbed out, the crew lost power on No 3 engine. Believing that the problem was an engine fire, the captain turned back towards Schipol for an emergency landing. What the crew did not realise was that they had truly lost the starboard inner engine. It had fallen away, probably because of metal fatigue in the engine attachment mountings. While separating, the errant engine hit its neighbouring No 4 engine, breaking it off also. The loss of these two engines, plus additional damage to the right wing, led to total loss of control. The 747 yawed and rolled into a spiral descent which only ended when the great aircraft ploughed into a block of flats near Biljemeer, Amsterdam, killing the crew of four and over 60 people on the ground.
PA News

Abbreviations & Glossary

AAIB	Air Accidents Investigation Branch
ACM	Air Chief Marshal
ADF	Automatic direction finder. Basic navigational instrument, used in conjunction with a ground-based radio beacon.
AGL	Above ground level
Airway	Designated air route, usually defined by ground-based radio beacons.
ATC	Air traffic control
Back course	Course flown along ILS beam in the reverse direction of the signal.
CAA	Civil Aviation Authority
CIC	Combat information centre
CVR	Cockpit voice recorder. Tape recorder that transcribes comments and audible actions of the flight crew.
Decision height	Specified altitude at which the crew must decide either to continue or to abandon a landing approach.
Decision speed (V_1)	Velocity at which the crew must decide to continue or to abandon a take-off.
Decision speed (V_2)	Minimum safe speed required in the air after an engine failure at V_1.
DME	Distance-measuring equipment. Equipment used to measure, in nautical miles, an aircraft's slant range from the beacon.
Dutch roll	Lateral oscillation of an aircraft involving both rolling and yawing.
Empennage	Tail section of an aircraft, including both horizontal and vertical stabilisers.
FAA	Federal Aviation Authority (US)
FDR	Flight data recorder. Transcribes vital flight performance information of an aircraft.
Feather	Adjustment of an aircraft's propeller to reduce drag following engine stoppage.
Fix	Geographical position determined by one or more ground navigation aids.
FL	Flight level. A level of constant atmospheric pressure related to a specific pressure datum (1,013.2mb). Each level is stated in three digits, eg FL250 represents a barometric altimeter indication of 25,000ft; FL255 an indication of 25,500ft.
Flight director	Instrument providing the pilot with pitch, roll and associated flight information.
GCA	Ground-controlled approach. Ground-based radar system in which the controller provides the pilot with vertical and horizontal guidance.
Holding pattern	'Racetrack' pattern to await landing.
ICAO	International Civil Aviation Organisation
IFR	Instrument flight rules. Guidelines used during a flight along an airway or specific route, usually while in radio contact with an air traffic control facility.
ILS	Instrument landing system. Standard landing aid comprising a glide slope beam for vertical and a localiser beam for lateral guidance.
IMC	Instrument meteorological conditions — ie flying on instruments only.
INS	Inertial navigation system. Capable of pinpointing an aircraft's position without reliance upon ground-based radio aids.
JFK	John F. Kennedy International airport, New York.
MDA	Minimum descent altitude. The lowest altitude, expressed in feet above mean sea level, to which descent is authorised on final approach during a standard instrument procedure where no electronic glide slope is provided.

Missed approach Abandonment of a landing approach, also termed 'overshoot'. A go-around' is a further attempt to land.

NDB Non-directional beacon. Enables the pilot of an aircraft equipped with direction finding equipment to determine a bearing to or from the transmitting radio beacon. The pilot can then 'home' on, or track to or from, the station.

nm Nautical mile. 1nm is about 1.15 land miles; 1kt (one knot) is 1nm/hr.

NTSB National Transportation Safety Board (US).

Octa Unit of measurement of cloud coverage, expressed in eighths.

Phonetic alphabet Alphabet used for clear and distinct verbal communication — Alpha, Bravo, Charlie, Delta, Echo, Foxtrot, Golf, Hotel, India, Juliet, Kilo, Lima, Mike, November, Oscar, Papa, Quebec, Romeo, Sierra, Tango, Uniform, Victor, Whiskey, X-Ray, Yankee, Zulu. Also used as a shorthand when referring to aircraft, so G-ARPI becomes 'Papa India'.

psi lb/sq in.

QNH Mean sea level altimeter pressure setting.

Rotation speed (V_R) Speed for raising aircraft's nose gear off the ground during take-off.

rpm Revs per minute

R/T Radio telephony

Runway number Figure at the start of a runway that denotes its compass heading when multiplied by 10 — eg 08/26 is a runway whose orientation is a heading of 080°(virtually due east) one way and 260°(virtually due west) the other.

RVR Runway visual range. Horizontal distance visible when looking down a runway centreline.

SID Standard instrument departure

Slat Movable portion of the leading edge providing additional lift to an aircraft's wings.

Spoiler Hinged surface on the upper surface of a wing designed to reduce lift.

Squawk Manipulation of a transponder to assist in identifying an aircraft.

SRB Solid-fuel rocket boosters

SSR Secondary surveillance radar — see Transponder

Stall Breakdown in the airflow around an airfoil, leading to a loss of lift.

Stick-shaker Mechanism designed literally to shake the control stick or wheel to warn of an impending stall.

TACAN Tactical air navigation. Ground-based UHF air navigation aid which provides suitably equipped aircraft with continuous bearing and distance information.

TAS True air speed

TCA Terminal control area

Time Reported in this book by means of the 24hr clock. So 16.16hrs is 4.16pm, and 16.16:23 is 23 seconds past 4.16pm.

Transponder Otherwise known as secondary surveillance radar (SSR). Enables each suitably equipped aircraft to superimpose its individual designator on the radar return painted on a ground controller's screen. In enabling controllers to differentiate between 'blips' with accuracy, it confers a higher degree of safety in a crowded and high-speed environment such as an airway. If an emergency occurs, the pilot of an aircraft fitted with a transponder 'squawks' Emergency Code 7700 to alert the ground agency.

VFR Visual flight rules. Guidance of an aircraft by the pilot involving traffic separation by sight, usually independent of an air traffic control facility.

VMC Visual meteorological conditions.

VOR VHF omnidirectional range. A ground-based electronic navigation aid transmitting VHF signals 360° in azimuth, orientated from magnetic north.

VSI Vertical speed indicator.

Wind shear Currents representing a significant change, in terms of direction or speed, from the general airflow.

Introduction

This is my third book on aircraft accidents to be published by Ian Allan Ltd. The first, *Crash!* (1991), was devoted to military mishaps, while *Disaster in the Air* (1992) concentrated on civil aviation. The demand for a third reflects abiding popular interest in aircraft safety. This interest is understandable. In 1944, when the Allies convened in Chicago to plan for postwar civil aviation, flying was a luxury reserved for a few. Only nine million passengers were carried by civil airlines in 1945; by 1993, that figure had expanded to 1.17 billion. Such massive growth is clear to see in peak-time throngs at London Heathrow or any other major international airport. At any given hour of the day, some 10,000 civil aircraft are in flight around the globe; to misquote the philosopher Rousseau, 'Man was born free but everywhere he is in planes'. Passenger traffic is forecast to rise to 1.8 billion by 2001, which means that we can expect to see bigger and bigger aeroplanes. Fifty years ago, the standard DC-3 could carry only 28 passengers. The first generation of pressurised aircraft was larger, but it took until 1956 before commercial aviation suffered its first disaster to claim more than 100 lives, and even that was only when two transports collided. By the beginning of 1970, there had been 18 accidents claiming more than 100 lives, but the record death toll of 134 over New York City would not be exceeded until the introduction of jumbo jets.

The first Boeing 747 — capable of carrying 66 first class and 308 tourist class passengers — entered service in 1970. Along with the Lockheed TriStar and McDonnell Douglas DC-10, these 'wide bodies' brought about unparalleled standards of safety and performance; the first 747 crash would not occur until nearly five years and more than two million flying hours into its service life. But when a wide body goes down, the world notices because of the numbers on board. And the world will take an even keener interest when the next generation of airliners, planned to accommodate 800 passengers, arrives at the departure gate.

Notwithstanding the massive growth in air travel over the last 50 years, let me say from the outset that flying is safer now than it has ever been. We have reached the stage where it is nearly 20 times safer to get airborne in an airliner than it is to drive to the airport in the first place. More people die in England and Wales from falling down stairs (568 in a typical year) than from air crashes.

Consequently, this is not a 'shock, horror' book. Rather, it tries to explain the background behind why airlines do some of the things — before, during and after flight — that they do today. Furthermore, I have flown enough aircraft types to know that flight safety is more than a litany of conspiracy theories. Anyone can come up with simplistic solutions. In 1982, Mr Dale Lowdermilk formed the National Organisation Taunting Safety and Fairness Everywhere in order to lampoon America's tendency to think that legislation can right every wrong. As a retired air traffic controller, Mr Lowdermilk's deepest concerns are with air safety. As such, he believes that aircraft should taxi rather than fly to their destinations. If they must fly, let them do so only one at a time. Passengers would be required to fly naked: this would stop hijackings, as there would be no place to hide bombs or guns. None of this would be convenient but it would meet the remit of those fanatics who put safety above all else. The US writer and critic H. L. Mencken put it best when he said, 'For every complex and difficult problem, there is a solution that is simple, straightforward, and wrong!'

The only way to guarantee never having an aircraft accident is never to fly in the first place. So, in dealing with past accidents, I have tried to pass on food for thought rather than pious platitudes.

Finally, as it is so easy to be clever with hindsight, I hope that I do not give offence to aviators involved in accidents, or to their friends and relatives. I regard all those who had the misfortune to be involved in any flying accidents as colleagues who would wish successive generations to learn from their experiences. *Flights to Disaster* is certainly not written to point the finger at anyone — I have enough pilot hours under my belt to admit that, sometimes, there but for the Grace of God went I. If this book, like *Crash!* and *Disaster in the Air* before it, stops anyone from thinking that aircraft cannot bite, and stimulates thought as to what to do should that awful moment occur, it will have done its job.

Andrew Brookes

1 As Luck Would Have It

Contrary to popular belief, Adolf Hitler spent very little time in Berlin after the start of World War 2. As self-proclaimed 'Supreme Warlord', he directed the German war effort from various Führer headquarters, the largest of these command posts being the reinforced concrete *Wolfsschanze* (Wolf's Redoubt) in East Prussia. By 1941 such was Hitler's authority that everyone, including ministers and military chiefs of staff, were 'invited guests' to the Führer's headquarters. The Wolf's Redoubt was served by Rastenburg airfield located on a plateau some 6km to the southwest. It was a reflection of Hitler's schedule after the invasion of Russia that his personal air squadron was based at Rastenburg. Only the favoured few got to fly on the Führer's aircraft and one of these was Sepp Dietrich, former leader of Hitler's bodyguard and, in January 1942, commander of an SS tank corps hard-pressed by the Russians in the southern Ukraine. On 30 January Dietrich flew to the industrial city of Dnepropetrovsk, taking along with him Albert Speer, Hitler's personal architect. Speer designed some of the most grandiose monuments to the Nazi regime, but on this occasion his staff were engaged on the more mundane work of organising the repair of railways in southern Russia.

According to his original schedule, Speer should have been back in Berlin by the evening of 8 February. He had completed his inspections two days earlier and as a Russian tank battalion was closing in on Dnepropetrovsk, Speer got himself on to a hospital train leaving that night. But a few hours out of the city, the train was stopped by 10ft snowdrifts — it returned to the city, arriving at dawn on 7 February. Two hours later, the pilot of Sepp Dietrich's aircraft, retired Lufthansa Capt Hermann Nein, offered Speer a lift at least part of the way home.

Around 11.00hrs, Capt Nein managed to take off from a runway poorly cleared of drifts, bound for Rastenburg. Huddled in the twin-engined Heinkel He111 bomber refitted as a passenger aircraft, the tall Speer had plenty of time to reflect on the value of air travel as he flew over the uninviting Pripet marshes. It was nightfall by the time the Heinkel landed at Rastenburg, which seemed quite beautiful after Speer's few days in Russia.

Having arrived unexpectedly at the Wolf's Redoubt, Speer put out feelers to see if he could secure an audience with the Führer. But Hitler was tied up in conference with Dr Fritz Todt, builder of the *Autobahnen* (motorways) and Siegfried Line; now Minister of Armaments and Munitions whose subordinates were undertaking massive communications works in support of the Russian offensive. It was late into the night before Todt emerged, strained and fatigued from a long and trying discussion. He wore a depressed air, quite possibly because even at this stage he was telling Hitler that the war in Russia was unwinnable. During the course of a rather lame conversation, Todt mentioned to Speer that he was flying to Munich the next day in his

personal He111. As the aircraft would stage through Berlin with an unoccupied seat aboard, Dr Todt said that he would be glad to take Speer along. 'I was relieved not to have to make that long trip by rail', recounted Speer. 'We agreed to fly at an early hour and Dr Todt bade me goodnight.'

An adjutant then came in and asked Speer to join Hitler. It was one o'clock in the morning but Hitler was a night person and Speer was used to such nocturnal conferences. 'When I finally left Hitler at three o'clock in the morning, I sent word that I would not be flying with Dr Todt. The plane was to start five hours later, I was worn out and wanted only to have a decent sleep.'

Next morning, the shrill clang of the telephone startled Speer out of a deep slumber. Dr Brandt, Hitler's personal physician and Speer's friend, reported excitedly that Dr Todt's plane had crashed on take-off around 07.55hrs, and that the Minister for Armaments and Munitions, not to mention inspector-general of roads, water and power, had been killed. A Reich Air Ministry Commission of Inquiry, chaired by Field Marshal Erhard Milch, was set up to investigate the cause of the accident. The SS was also involved, doubtless in search of sabotage or other skulduggery, but neither Air Ministry nor SS could provide any firm explanation. In the absence of hard facts, many conspiracy theories have been put forward for this accident: these included Hitler's desire to rid himself of an articulate opponent of the war with Russia, a successful strike by Allied security services, or activation of the 'self-destruct mechanism located between the pilots' seats'. The cause was probably far more prosaic. Being a tri-motor Junkers Ju52 man himself, Hitler issued an edict forbidding any of his top people from using twin-engined aircraft. But Todt would have none of this, and he obtained a new twin-engined He111 at the end of 1941 for his travels. Concerned about Dr Todt's safety in this relatively new aeroplane, Lt-Col Klaus von Below, Hitler's adjutant as well as the ranking Luftwaffe officer at Rastenburg, arranged for the He111 to be taken up for a test flight before Dr Todt's departure. It was perfectly serviceable and being just a standard He111 communications variant, there was no self-destruct mechanism aboard.

After being told of Dr Todt's death by the flight captain of the Führer courier squadron, von Below hurried over to the airfield. 'I found nothing there except still-smoking wreckage; all the occupants of the flight were dead. After Hitler got up, I reported the crash. He was startled and was quiet for a long time. Then he asked me what the cause was, but I had no explanation.' Yet being a shrewd airman, von Below got as near to the truth as anyone while he gazed over Rastenburg airfield on that fateful morning. 'The weather was not good, the sky and the snow-covered ground was grey, the line between conceivably invisible. I suspected human error by the pilot, who was not yet sufficiently familiar with the plane in difficult weather conditions.'

But whatever the cause of the accident, from that moment, on Albert Speer's whole world was changed. Although still only 36 years old and lacking in any experience of large-scale industry, Speer was appointed to succeed Todt within five hours of the accident; almost at once he began to demonstrate the most remarkable talent for the administration of war industry. When Todt died, he controlled armament and construction programmes involving 2.6 million workers. By late 1944, the total workforce for which Speer (by then in charge of all civilian and military production of the Great Reich) was responsible, totalled 28 million. Despite the rising tide of Allied bombing attacks on German industrial facilities, the gifted Speer was able, through a policy of rationalisation and dispersal combined with far greater latitude than Todt ever possessed, to make output rise every month until September 1944, and to maintain a creditable flow of war *matériel* right to the very end. It is no exaggeration to say that if any one individual kept the war in Europe going for as long as it did — and some would say extended it by up to 12 months — that person was Albert Speer. Yet if fate had dealt the cards ever so slightly differently at Rastenburg in February 1942, Albert Speer would have died with Dr Todt. Which only goes to show that Lady Luck does not just snuggle up to the good guys. As with life in general, there is nothing intrinsically fair about flying.

The only thing that surpasses missing a fated flight is to give up your terminal ticket to someone else. In late 1958, the rock'n'roll star Buddy Holly had split with his producer and the Crickets — disbanding the team that had made such great music during 1957 and 1958. Since the parting of the ways, Holly had been waiting for finalised accounts of the royalties owing to him, and by January 1959 the singer's

Opposite: *Fritz Todt (left), creator of the Autobahnen and Hitler's Minister of Armaments from 1940, with Albert Speer who was to succeed him after the He111 air crash in February 1942.*

existing funds were becoming drained, a situation not eased by the fact that Mrs Holly was confirmed to be pregnant. Needs must when money drives, and in the middle of January Holly reluctantly agreed to leave his wife in New York while he joined a 'Winter Dance Party' package of rock'n'roll musicians being set up to tour the US Midwest.

The party consisted of five acts: Holly, Ritchie Valens, J. P. Richardson (known as the 'Big Bopper'), Frankie Sardo and a group called Dion and the Belmonts. They were to tour towns in Wisconsin, Iowa, Minnesota and North Dakota, and there are few more bitterly cold places than these in February. Dion Di Mucci, leader of Dion and the Belmonts, was used to cold winters in his native Bronx, but he recalled those early February days as 'just about as bad as any I can remember. It was so cold that one of the guys actually got frostbite and had to be left behind in hospital. We travelled by bus, sometimes four or five hundred miles between dates. There was never any heating in them and you had to fool around to keep warm. Getting some sleep was a real problem.'

It was only the enthusiasm of the young audiences that kept the entertainers going before they reached the small town of Clear Lake in northern Iowa on Monday, 2 February. The atmosphere was as charged as ever in the Surf Ballroom, quaintly named as it was 1,000 miles from any ocean, but after the show Buddy Holly told the other artists that he had chartered a light aircraft at Mason City — some 10 miles from Clear Lake — to fly to Fargo, North Dakota, near the next venue on their agenda. His lead guitarist, Tommy Allsup, and his bass guitarist, the then unknown Waylon Jennings, were going with him.

In the event, Ritchie Valens and J. P. Richardson talked the backing musicians into letting them have their aeroplane seats — in Tommy Allsup's case, following the flip of a coin. Di Mucci knew none of this until he arrived at Fargo to be told that there had been a crash. The basic facts, as determined by the US Civil Aeronautics Board (CAB) investigation, were that Buddy Holly chartered a four-seat, single-engined Beechcraft Bonanza (N3794N) from the Dwyer Flying Service to fly to Fargo. The aircraft, flown by a comparatively inexperienced local pilot, 21-year-old Roger Peterson, took off around 00.55hrs on 3 February. It crashed barely five miles northwest of Mason City municipal airport on a lightly snow-covered cornfield belonging to Albert Juhl. The pilot's body was found in the wreckage while those of the three singers were close by. The CAB report attributed the main cause of the accident, which coincidentally occurred on Holly's 13th tour, to pilot error when operating in sub-zero conditions while heavy snow was falling.

A whole host of famous people have died in air crashes over the years including UN Secretary General Dag Hammarskjöld in 1961, former undefeated heavyweight boxing champion Rocky Marciano when the light aircraft in which he was travelling crashed just south of Newton airport, Des Moines, Iowa, in 1969 and former world champion racing driver Graham Hill who was killed when his twin-engined aircraft crashed in fog near Elstree airport in 1975. The list could go on and on, though one has to be careful about the last flight of Glenn Miller, the great wartime band leader booked to broadcast to America from Paris on Christmas Day, 1944. His plan was to travel on ahead of his

Opposite: *A four-seat Beechcraft Bonanza similar to that in which Buddy Holly died.*

orchestra, but pretty convincing evidence has come to light to cast grave doubts on the legend that he died on 15 December 1944 in single-engined UC-64A Noorduyn Norseman (44-70285) which came down in the English Channel. In the end we may well find that Glenn Miller died in circumstances that had little to do with aviation, which is perhaps a long-winded way of making the point that the only way to guarantee avoiding an accident in the air is never to get on board an aeroplane in the first place!

Maj Glenn Miller never pretended to be anything other than a passenger in the air, but others were not so firm in sticking to what they did best. A prime example was James Travis Reeves, better known to millions of country music fans as 'Gentleman Jim Reeves'. From humble beginnings as a journeyman music-maker, Reeves changed his approach to singing in 1955 by pitching his voice lower and singing close to the microphone, thereby creating a warm ballad style which made him enormously popular around the world, not least in South Africa. Reeves did not like flying but after being a passenger in a South African aircraft which developed engine trouble, he obtained his own private pilot's licence. Like many other private pilots, Reeves was qualified only for fair weather operations outside controlled airspace, but on 31 July 1964 he and his pianist/manager died when their single-engined aircraft ran into difficulties during a storm and crashed into dense woods outside Nashville, Tennessee. It took two days of searching by 500 people before their bodies were found.

Although Reeves continued to have posthumous hits with such ironic titles as 'The World Is Not My Home' and the self-penned 'Is It Really Over?', his sad loss in addition to all the other famous air crash casualties underlines two basic truths about aviation. First, although the winds tend to favour the best navigator, aviation in general is no respecter of persons. And second, flying is a serious business. Small aircraft can kill just as effectively as big ones, and wise aircrew respect their personal limitations and stick to them. Just because you can sing like an angel does not mean that you can fly like one.

2 Prepare for Take-Off

When the British set up the Aeronautical Inspection Directorate at Farnborough in 1913, one of its responsibilities was the investigation of military aircraft accidents. This role passed to the Accidents Investigation Branch (AIB) when it was founded in 1915, with an independent 'Inspector of Accidents' reporting to the Director General of Military Aeronautics in the War Office. After World War 1, when commercial companies started operating between London and the Continent, it became clear that some organisation needed to take responsibility for the safety of fare-paying civilian passengers. The Department of Civil Aviation was therefore set up within the Air Ministry, and the AIB joined it in 1920 with a brief to investigate both civil and military flying accidents.

This arrangement sufficed for over a quarter of a century because it met the needs of the times — after all, there were only 26 flights across the Atlantic during the whole of 1940. But at the end of World War 2 a separate Ministry of Civil Aviation was established to co-ordinate the expected growth in commercial air travel. The AIB was transferred to it in 1946.

Below: *A Railway Air Services' DC-3 Dakota after it failed to get airborne in bad weather from Northolt, West London, on 19 November 1946.*

British civil air accidents are now the responsibility of the Department of Transport's Air Accidents Investigation Branch (AAIB), but the branch's objectives remain as they ever were: 'To determine the circumstances and causes of an accident with a view to the preservation of life and the avoidance of accidents in the future.' In the interests of impartiality, the AAIB does not apportion blame or liability; such post-inquiry considerations are left to the Civil Aviation Authority (CAA) or the airline operator concerned.

Other major nations work along similar lines. When a US-registered aircraft is involved in an accident, the National Transportation Safety Board (NTSB) — successor to the Civil Aeronatics Board — sends out an investigating team to ferret out the cause or causes. The NTSB then makes recommendations to the aircraft operators for action, or to the Federal Aviation Authority (FAA) for permanent regulations, to lessen the probability of repetition.

Statistically, the most dangerous time in an aircraft's life is during take-off. This is not just because of the proximity of the ground — it is also the time when aircraft speed is building up and, if anything goes wrong, there is next to nothing spare to convert to precious height. The first air tragedy to claim over 100 lives happened on 18 June 1953 when a USAF C-124 Globemaster II crashed on take-off from Tachikawa air base near Tokyo, Japan, killing all 129 people on board. Ten years later another take-off accident took place in Switzerland.

Wednesday, 4 September 1963, was planned to be a big day for the tiny Swiss hamlet of Humlikon located to the north of Zurich. Forty three of its leading citizens — one quarter of the population including its mayor and town councillors — were leaving early in the morning for Geneva to attend an agricultural show. This event was of great interest to the farmers and their wives from this close-knit rural community, but it promised to be just as exciting to travel as to arrive. Instead of a tortuous 300km train journey through the mountains, the good burghers of Humlikon were booked to fly from Zurich's Kloten airport to Geneva on one of Swissair's new Sud-Aviation Caravelle twin jets.

Swissair Flight 306 was Caravelle III HB-ICV scheduled to depart at 07.00hrs. In command was Capt Eugen Hohli, who had been with Swissair for a decade, assisted by First Officer Rudolph Widmel. Although the day was fine and cloudless on the mountains, shallow morning fog lay in the

valleys and Kloten was still closed to operations as departure time approached. Nothing daunted, Capt Hohli decided to have a look at conditions on the runway and so the passengers — all bar six of whom were Swiss — were ushered aboard at the normal time.

Once engines were started, Flight 306 asked for the surface visibility on Runway 16/34. ATC replied that visibility was only 60m from the 16 threshold but that it was 210m at the 34 end. The crew then asked for a Follow-Me truck to guide them to the 34 threshold plus approval to 'taxi on the runway — on 34 — and back again to have a look around'.

A truck duly led the slowly taxying Caravelle out to 34 but the fog was so dense that the guide vehicle driver eventually lost his way, mistakenly taking the airliner down Taxiway 4 instead of Taxiway 5; consequently, HB-ICV joined the runway about 400m from the threshold rather than at the threshold itself. The crew turned on to the runway and, with engines thrusting loudly, began taxying slowly in the direction of take-off. After covering about 1,400m, the Caravelle was turned through 180° before taxying slowly back to the 34 threshold. Workmen close to the runway noticed that the noise from the Rolls-Royce Avon engines was a good deal louder than normal for a taxying Caravelle.

At 07.09hrs the flight crew reported that the fog appeared to be lying in banks with varying visibility, and

Opposite:
A Douglas C-124 Globemaster II strategic transport, designed to carry heavy vehicles or 200 fully-equipped troops, over San Francisco. The loss of a Globemaster such as this on 18 June 1953 was the first air disaster to claim over 100 lives.

Above:
A Caravelle, similar to that which crashed in Swissair colours shortly after take-off from Zurich on 4 September 1963.

that the efflux from their engines had some effect in clearing the fog from the runway. Three minutes later, Flight 306 requested clearance to take off. ATC responded promptly and the Caravelle took off normally. Shortly afterwards the crew reported that they were in the clear above the fog bank at 1,700ft, climbing to cruising level. They were handed over to Zurich departure control but, eight minutes later, 306 transmitted a Mayday. The pilot's voice then became almost incoherent, and when air traffic sought details of the emergency, he gasped with desperation, 'No more...' Silence followed, and all further attempts to communicate with the Caravelle went unanswered.

Around this time, a farmer working high on a valley side sighted the Caravelle. It appeared to be flying normally in a clear sky when whitish smoke suddenly started to stream from the port side. As the farmer continued to watch, flames then burst out along the port side of the fuselage. The airliner flew on for a time, seemingly unaffected, but then it entered a gradual turn to the left before losing height. Finally HB-ICV nosed over into a dive and plunged from sight back into the fog.

At Durrenasch, 15 miles southwest of Zurich, local businessman Heinrich Lienhard was breakfasting in the kitchen of his two-storey house when he became aware of the sound of an approaching jet aircraft. But instead of passing by, the noise increased until it became ear-shattering. Seconds later, a tremendous explosion rocked the house like an earthquake; debris sliced into the roof, carrying away tiles and beams. Herr Lienhard's house was one of several buildings that were damaged by debris scattered over half a square kilometre as HB-ICV gouged a crater 20m long by 10m deep into a potato field on the edge of the village. Heinrich Lienhard was unhurt, but not a single body was recovered from the impact crater or wreckage trail stretching beyond it. Of the Caravelle's capacity load of 74 passengers and six crew, next to nothing remained. It was the worst disaster in Swiss aviation history.

It was clear that the Caravelle had caught fire in the air, but for what reason? Wreckage in the impact crater was pretty mangled but parts of the

aircraft, from the port wing and rear fuselage, were later found on the ground under the last 6nm of HB-ICV's flightpath. Fragments of the outer rim of its No 4 landing wheel (the inner rear of the port undercarriage bogie) were also found near the threshold of Zurich's Runway 34. Close to it on the runway were a blow-out stain, an earthing cable and its attachments, and traces of burnt hydraulic oil. The blow-out stain lay between the tracks of the two pairs of wheels of the port main undercarriage; beyond this point, the wheel tracks were covered in hydraulic oil.

Swiss accident investigators deduced that the wheel flange of No 4 landing wheel split as the Caravelle was being turned around the runway threshold to line up for take-off; this caused the corresponding tyre to explode. The condition of the wheel's brake, recovered after much effort from the crash site, showed that the mechanism had been grossly overheated by the prolonged manoeuvres along the runway before take-off. Tests showed that such overheating would lead to a wheel fracture similar to that which occurred in HB-ICV's No 4 wheel.

After the Caravelle got airborne, eyewitnesses saw a trail of white smoke that suddenly became a major fire in the vicinity of the port undercarriage. This raised two alternative scenarios. First, a similarly stressed No 3 wheel (rear outer on the port bogie, adjacent to No 4) ruptured after the Caravelle got airborne. Tests showed that wheel rims of the type fitted to the Caravelle did not reach their maximum temperature until a few minutes after the braking effort that caused the overheating. It was therefore possible that No 3 wheel rim had burst after take-off, causing damage to fuel lines and a separate fire which eventually burnt its way out of the undercarriage housing.

However, as the runway bore witness to the fact that hydraulic oil had escaped from the braking system immediately after the blow-out and this oil was burning as the take-off began, it was more likely that the blaze in No 4 wheel continued to burn after gear retraction. This then ignited No 3, causing its failure. This fire could be expected to have spread to the other two wheels of the port bogie, and eventually out of the undercarriage housing.

As they broke out on top of the fog, passengers and crew were blissfully unaware of what was developing beneath them. But once the fire took hold around an altitude of 9,000ft, the Caravelle veered left and then entered a steep dive. This was probably induced by a loss of rigidity in the port wing structure due to the fire, loss of hydraulic power for the control system, and structural damage to the port tailplane.

There could be no doubt that the fire which eventually orphaned 43 children from Humlikon was brought about simply by the overheating of brakes before take-off. In essence, the crew had used high power settings to try and disperse the fog while deliberately holding the aircraft to a low taxying speed on the brakes. Two Avon engines, even if they were not pushing out their collective maximum 22,800lb thrust, took some holding. Leaving aside the wisdom or otherwise of the fog-burning technique employed that day at Zurich airport, by the time Flight 306 took off its brakes would have become so overheated as to be totally useless for their primary purpose — emergency braking. It is particularly galling when a flight crew misuses

Opposite: *The Swiss village of Durrenasch, showing the impact crater and damaged adjacent houses after Flight 306 impacted.*

a safety device to the extent that it kills them. It is vitally important that pilots allow time for brakes to cool between abandoning a take-off for whatever reason (nearly always with heavy braking) and beginning another take-off run. It never pays to degrade any margin of safety.

But take-off disasters rarely stem from one cause — as with most flying accidents, they usually result from a combination of factors. On 21 June 1974, a Dan-Air Boeing 727 (G-BAEF) was scheduled to make a charter flight from Luton to Corfu with 126 passengers and eight crew. On boarding, the flight engineer mentioned that the loadsheet showed the airliner to be 125lb overweight for a take-off on Runway 26 in zero wind. The captain replied to the effect that there was some headwind component that would take care of this.

At 06.37hrs the pilot was told that his proposed take-off time 'slot' was 06.54hrs plus or minus 4min; he replied that he could comply. At 06.40hrs another aircraft was told that the duty runway was 26, the wind calm and runway 08 was available; the pilot of that aircraft chose to take-off on 08.

At 06.49hrs the pilot of G-BAEF requested taxi clearance, and he was cleared to the holding point for runway 26. In accordance with standard ATC practice, this clearance did not include the surface wind but the captain subsequently stated that he understood a surface wind of 230°/8kt was given at this time. 'Before take-off checks' were completed during taxying, and the airliner was cleared to backtrack down the active runway to the take-off position. While making a 180° turn around the dumb-bell at the end of the runway, the crew received take-off clearance plus the surface wind as 150°/less than 5kt. Final 'runway items' of the checklist were carried out during the turn, and an immediate rolling take-off was started from a position which the captain later estimated was about 200ft in from the beginning of the runway. Take-off began shortly after 06.54hrs, flown by the first officer from the right-hand seat.

Although the ground run appeared to be normal, witnesses noted that the nosewheel was not lifted until later than usual, at a position just before or abeam of the runway visual range observation point (just over 6,000ft from the beginning of Runway 26). The wheels remained on the ground, and the airliner continued in a partially rotated attitude past the end of the runway; it appeared to get airborne just at the end of the stopway. One observer said it seemed 'that the aircraft remained level while the runway dropped away from it'. The 727 was then seen low in the valley to the west of the airport whence it climbed away, slowly at first but then normally after attaining a fully rotated position.

Both the ATC local controller and fireman in the airport fire station watch room were convinced that G-BAEF had struck the ILS localiser aerial, situated 550ft past the runway end. Inspection of the aerial confirmed this to be so and that a number of Runway 08 approach lights had also been damaged. None of this had been noticed on the flightdeck where the take-off had appeared to be completely normal. However, as the 727 passed 3,000ft accelerating in the climb, the first officer reported that excessive aileron was required to correct a heavy left-wing condition. The commander took control, and he found that 4-5° of right-wing-down aileron trim was required to keep the aircraft laterally level. It was then that Luton Tower informed the crew that their aircraft had probably struck the localiser aerial. After fuel was dumped over the sea to the south of Worthing, G-BAEF was

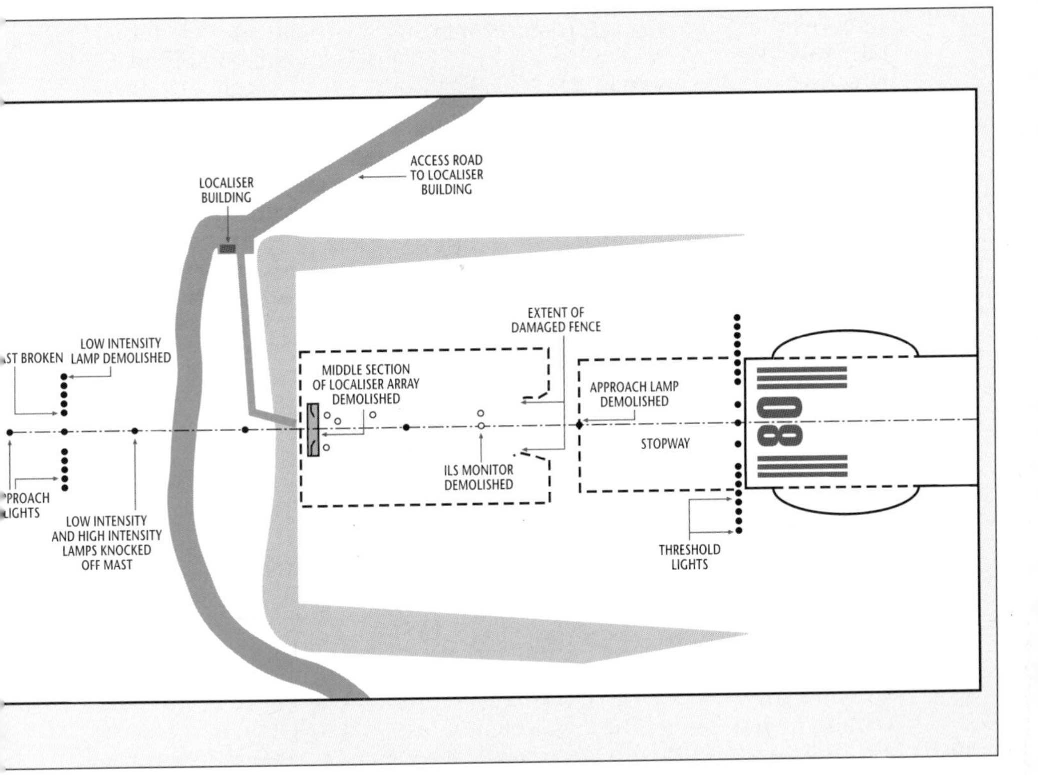

landed without further incident at Gatwick. It was found that both main landing gear wing doors had been wrenched off their forward hinges, the tail skid was broken and a 3-4ft gash had been made along the centre keel of the fuselage at the forward end of the rear cargo compartment.

It was apparent from flight data recorder evidence and eyewitnesses that G-BAEF's take-off rotation was initiated and achieved much further along the runway than the scheduled position. From the damage sustained by both the aircraft and the ground installations at Luton airport, it was also apparent that the 727 was dangerously low during its initial flightpath after take-off.

According to the operations manual, G-BAEF was overweight for Runway 26 in zero wind; it was even worse placed for a surface wind of 150°/less than 5kt, which gave a slight tailwind component. When the true surface wind was passed during taxying, consideration should have been given to using Runway 08, partly because of the slight beneficial headwind but also because the overall slope would have been more favourable. It is possible that the closeness of the take-off time slot, which expired at 06.58hrs, played some part in the decision to remain with 26.

However, such an oversight should only have extended the take-off run required from 4,120ft to 4,580ft. If 300ft were added to offset the distance lost during the 180° pre-take-off turn on the dumb-bell, the 727 would have used 4,780ft from brakes off to scheduled V_R. But the flying pilot

Above: *Installations at the end of Luton's Runway 26 which were given a glancing blow by a Dan-Air Boeing 727 on 21 June 1974.*

delayed the point of rotation until some 8sec above the scheduled V_R, which added a further 700ft (5,480ft). Eyewitnesses implied that it might have been even later than this (6,067ft).

Nevertheless, Runway 26 plus 200ft of stopway provided a total length of 7,287ft, so G-BAEF still had room to spare. But the final 1,940ft of 26 plus stopway had a down slope of 0.57%. As the 727 had not been fully rotated at lift-off, it flew approximately parallel to the slope after it left the ground, a conclusion borne out by the fact that it struck the innermost approach light for Runway 08 which projected about one foot above the stopway surface. The slow rate of rotation continued whilst the airliner followed a flightpath which eventually took it below the down-sloping portion of the clearway plane to a point about 1,000ft beyond the end of the runway. During this time it struck the localiser and some of the Runway 08 approach lights. It was not until about 13sec after start of rotation that G-BAEF was fully rotated to an attitude of about 13°, sufficient to initiate a positive climb. Over that period, it was fortunate that there was a valley to the west of the airport.

It was the cumulative effect of the eroded take-off run, the delay in starting rotation and the very slow rate of rotation that so nearly did for G-BAEF that day. Any of these factors would have been individually acceptable, but not them all. The 33-year-old first pilot had taken off from Luton on 15 previous occasions, but this did not prevent him (and his 51-year-old commander with 15,818 flying hours) from completely misjudging the situation. Flying safely must always be a positive business, but just in case it is not your day, it is worth remembering that one of the most useless things in aviation is the runway behind you.

You always need your wits about you when in control of an aeroplane. On 12 May 1995, a Lufthansa Airbus A300 arrived at the back of the holding point queue for London Heathrow's Runway 09R. The standard instrument departure (SID) altitudes for some aircraft had been amended because of a special flight in the area; these aircraft were instructed by ATC to maintain 3,000ft after take-off until further advised. It was not known by the flight crew how long the temporary restriction to the SID was going to continue, but those ahead of the Lufthansa Airbus were given the SID amendment before being instructed to line-up for take-off clearance.

The Lufthansa crew was not given an amended SID when the aircraft lined up on the runway, so an altitude restriction was not expected. The crew therefore believed that the next transmission would be the take-off clearance, an expectation reinforced by the sight of preceding airliners lining up and almost immediately lifting off. The crew then received an amended SID.

It was obvious from the ATC recording that the Lufthansa crew was mistaken in its interpretation of the air traffic message. This was clearly an amended SID and although English was not the pilots' first language, their command of the idiom was near perfect. Yet, having primed themselves mentally for take-off clearance, that is what they thought they heard.

The price of this delusion could have been very high. Several seconds later, while 150 tons of fully-laden Airbus was accelerating past 100kt, the crew saw traffic crossing the runway ahead. As there was ample room, the crew decided to continue the take-off; in doing so, they passed above two vehicles and a taxying Boeing 747 which had just crossed the runway.

The Lufthansa crew commented to ATC that to be cleared for take-off with ground traffic crossing was not a good idea, only to be told that take-off clearance had not been given!

The problem of 'expectation' influencing what is heard on R/T is well recognised. In this instance, the crew was probably right to continue and rely on 120,000lb of engine thrust to keep the Airbus out of trouble. But raw power can only do so much. On 16 August 1987, Northwest Airlines Flight 255 took off from Runway 03 at Detroit Metropolitan Wayne County airport, bound for Phoenix, Arizona. The McDonnell Douglas DC-9 Super 82 (N312RC) had been airborne for only 14sec when, at a height of 50ft above the ground, the airliner's port wing struck a lamp standard located about half a mile beyond the end of the runway. Its undercarriage was still in the process of retracting when the DC-9 clipped other lamp standards and the roof of a building. Some 18ft of outer port wing having been torn off in the initial impact, the airliner rolled through 90° to the left and slammed to earth, disintegrating and bursting into flames. Wreckage was scattered along a road, and under a railway and two highway overpasses. A total of 156 people died that day in Romulus, Michigan, including the aircraft's six crew members and two occupants of vehicles hit by falling wreckage. The sole surviving passenger was a four-year-old girl.

Examination of the wreckage revealed no malfunction in the aircraft's engines, flying controls or avionics that could have accounted for the disaster. However, investigators did discover that the DC-9's flaps and leading edge slats were retracted at the time of impact.

The absence of extended flaps and slats would have severely limited N312RC's climb capability and increased its stalling speed. This was made clear by the airliner's relatively long take-off run and higher than normal pitch angle on becoming airborne. The stick-shaker stall warner was heard on the cockpit voice recorder (CVR) less than one second after lift-off. Once in the air, the DC-9 began rocking laterally, which the crew tried to control by deploying the spoilers. The 'Dutch roll' motions plus corrective action further degraded N132RC's performance.

Extension of flaps and slats were the first items on the taxi checklist, but playback of the CVR revealed that the two-man flight crew neither called for nor carried out this checklist. The first officer should have set the flaps and slats after the start of taxying, but around the time this should have been done he was noting a change in take-off runways. When the co-pilot had finished copying this automatic terminal information service message, the DC-9 had passed the point where extension would have been completed. This may have misled the co-pilot into believing that what he 'expected' to have happened had indeed done so.

According to Northwest operating procedures, the captain of Flight 255 should have initiated the checklist routine. In the event, he did not ask for the after-start, taxi or before take-off checks; instead, he delegated this responsibility to his first officer. The omission of the taxi checklist was regarded by the NTSB as being the primary cause of the disaster.

Super DC-9s such as N312RC were equipped with a sophisticated control aural warning system (CAWS) which, among other things, was designed to recognise the conditions which could engender a stall. If the thrust levers were moved and the system noted an improper flap/slat configuration, an electronic voice was supposed to intone 'flaps' and/or 'slats'. No such aural

warning was heard on the CVR recording after the loss of Flight 255, a failure attributed to a loss of electrical power to the CAWS. There was no way of determining the precise cause of this power loss, but the NTSB considered the power loss to be the principal contributing factor in the crash. Suffice to say that one can never rely on wondrous devices to offset human omission or haste. The words of Capt A. G. Lamplugh, written around 70 years ago, are just as relevant today: 'Aviation is itself not inherently dangerous, but to an even greater extent than the sea, it is terribly unforgiving of any carelessness, incapacity or neglect.'

But it is possible to get too far ahead of the game. On 20 July 1975, a Handley Page Herald (G-APWF) belonging to British Island Airways was making a scheduled passenger flight from Gatwick to Jersey. Whilst the captain was taxying the twin Dart-engined short/medium-range airliner, the co-pilot completed the checks. He asked the commander if the take-off was to be made with or without water methanol assistance (wet or dry), and when the answer came back 'dry', the co-pilot believed he selected 5° of flap which was obligatory when water methanol power was not being used. Shortly after the captain said that the take-off would be 'dry', the pre-take-off drill was interrupted when the co-pilot had to make a change of radio frequency. G-APWF was then cleared to line up on Runway 26 after the departure of a Boeing 707.

Once the 707 had taken off, the commander lined up G-APWF on the left-hand side of the runway to take advantage of the slight crosswind to disperse any residual wake turbulence from the preceding aircraft. As the Boeing passed 2,000ft, the Herald was cleared to go. Full take-off power was applied and verified by the co-pilot. As the speed built up through the standard calls of '80kt, V_1, V_R and V_2', the co-pilot was holding the throttle levers with his right hand and the throttle friction lever with his left. The commander said that he made a normal rotation to about 7° nose-up at V_R and, having observed a positive rate of climb of 70-100ft/min on the vertical speed indicator (VSI), he ordered the landing gear to be retracted. Seconds later, when the aircraft was about 15ft above the runway, it descended suddenly. The captain could do nothing to arrest it. At 09.00hrs the Herald settled back on to the runway before sliding to a stop, wheels up. All 41 passengers and four crew got out safely. The co-pilot told the subsequent AIB investigation that, on hearing the order to raise the gear, he had locked the throttle levers and pushed the 'up' selector button with his left thumb but that the switch would not depress. He then leaned to the left to obtain more leverage, pressed the 'up' button with his right thumb and the selector depressed without difficulty. He admitted that he did not observe the reading on the VSI before selecting gear 'up'. Whilst waiting to confirm that the undercarriage was fully retracted, he heard the commander make a remark which suggested that something was amiss. On looking up, he saw the aircraft close to the ground and not climbing.

No defect was found in the aircraft, its engines, flight instruments, flying controls or its warning and indicator systems. Nor were wing vortices from the preceding Boeing 707 a significant factor. The inquiry was left in no doubt that the events leading to the accident originated in the pilots' mistaken belief that the flaps had been extended to the take-off setting prior to entering the runway. But the flaps remained up throughout, an omission which was probably caused by the interruption in the pre-take-off drill sequence. Such interruptions are not unusual in aviation, and their disruptive effect is best prevented by proper use of checklists and careful cross-monitoring.

The flight data recorder (FDR) showed that although the take-off was made in a flaps-up configuration, the engine power and take-off airspeeds used were those appropriate to a wing flap setting of 5°. The airspeed at the time of rotation was therefore less than that required to sustain a safe initial climb. If the flaps had been extended to 5°, the aircraft would have been capable of climbing away safely at the speed and rotation angle achieved. As it was, the Herald went down. On the other hand, if the landing gear had not been retracting, it is unlikely that the touchdown would have resulted in anything more than a wheel bounce before G-APWF accelerated to an adequate take-off speed. The premature retraction was therefore identified as the primary causal factor in the accident. The co-pilot's first attempt at retraction was probably prevented by the weight-on switch which did its job while the main wheels were still on the ground. It is possible that, in his over-anxiety to comply with the commander's instruction, the co-pilot unconsciously exerted sufficient pressure to override the gear retraction baulk, but the more likely reason for his success was that the selection coincided with the main wheels lifting off.

The commander's order to retract the gear was undoubtedly premature. In all probability it was based on the knowledge that a positive rotation had been made, the observation of a small rate of climb and the confident anticipation of the normal climb resulting from a 5° flap setting. There were over 21,000 hours of flying experience on G-APWF's flightdeck that day, but they only made their presence felt in terms of over-familiarity with frequent routine activity.

Opposite: *Wreckage of Northwest Airlines Flight 255 scattered beneath road and railway bridges at Romulus, Michigan, after the DC-9 failed to stay airborne on 16 August 1987.*

Having said all that, there are some pitfalls that are next to impossible to avoid. Eastern was the first airline to put the short/medium-range Lockheed 188 Electra turboprop into scheduled service, and a year later — on 4 October 1960 — Eastern Air Lines Electra N5533 took off from Boston's Logan airport bound for Philadelphia. Designated Flight 375, the airliner had just got airborne around 17.40hrs from Runway 05 when it flew into a flock of starlings. A large number of birds was then ingested into three of the Electra's four Allison engines.

No 1 was shut down when its propeller autofeathered, No 2 flamed out completely and No 4 suffered a partial loss of power. The resulting asymmetric yaw to the left was exacerbated by the recovery of full power — 5,500 shaft horsepower (shp) — on the starboard outer before No 2 was relit. Meanwhile, the overall power loss caused N5533 to decelerate and then stall during a skidding left turn while continuing to yaw. After the port wing dropped and the nose pitched up, the airliner went into a spin before plummeting into the shallow water of the Boston Harbour inlet. The undercarriage was up and the flaps were extended as Flight 375 impacted some 500ft from the shore, killing 62 of those on board (59 passengers and three members of the flight crew). The 10 injured survivors included both stewardesses.

As Flight 375 was less than 150ft above the ground when it got mugged by starlings, the pilots had neither height nor speed to play with and were therefore unable to take any meaningful recovery action. The risks attendant on in-flight encounters with birds have been recognised since 3 April 1912 when C. P. Rogers, the first man to pilot an aircraft across the USA, took off from Long Beach, California, only to plunge to his death after running into a flock of seagulls. There was a time when it was thought that the distinctive sound made by turboprop engines attracted birds, but this was never proven. Bird strikes continued into the jet age with reported worldwide incidence currently being around 4,000 annually. And as an 8lb bird colliding with an aircraft travelling at 300mph produces forces calculated at 46,000lb, there is little practicable solace to be found in the wholesale beefing-up of airframes or systems.

As 66% of bird strikes occur at heights of less than 200ft above ground level and in the vicinity of airports, preventive measures such as scarecrows, falcons, noise-makers, marksmen and the prohibition of rubbish dumps near airports have been introduced with varying degrees of success. They will continue to play a crucial part in the reduction of risk to departing aircraft, but only a part. Migratory patterns or the propensity of birds to flock together around dawn and dusk are just two of the flight safety considerations that professional aircrew must hoist on board when it comes to being prepared to take to the air. Just as birds will always be with us, comprehensive take-off planning and meticulous execution are duties that good pilots must never avoid.

3 See and Be Seen

Despite the flimsy 'wire and string' nature of the first heavier-than-air flights, at least the Wright brothers had the skies to themselves. Once a second and subsequent pioneers got airborne, it would only be a matter of time before the first midair collision.

The first British commercial air service opened for business between London and Paris on 25 August 1919, and the first civil midair crash was not long in coming. On 7 April 1922, about 60 miles north of Paris, two airliners collided in midair, killing all seven people on board. Among the fatalities was a 16-year-old steward newly hired by Daimler Airways to serve coffee to passengers.

The accident was blamed on poor visibility. The pilot of the Daimler Airways DH18, heading for Paris from London, was flying at a few hundred feet through a valley near the town of Grandvilliers. His view was obscured by mist and the DH18's wide fuselage — the pilot's open cockpit sat behind the enclosed passenger cabin amidships — so he failed to see a Farman Goliath of the French Grands Express Aériens ahead of him. The Goliath pilot, flying from Paris to London, was following the same route at the same altitude in the opposite direction; he too failed to see the oncoming DH18, and the aeroplanes collided head-on. A newspaper of the day said that, 'the disaster will intensify calls for tighter control of air traffic. At the moment pilots are left a free hand in finding routes, and many follow the same ones.' By 10 July, French authorities established the first official one-way route from Paris to London: Le Bourget to Ecouen, along the Nationale 1 road to Abbeville, from there along the Paris–Calais railway line to Etaples and thence across the Channel via Oxted to Croydon. It sounds a charming relic of a bygone age, but it meshed the needs of the times with the technology of the age. Significant change would have to await a new era.

Wars are not wholly destructive. Survival is often the mother of invention, and just as the croissant came out of the 1683 siege of Vienna and improved health care from the Crimea, so the technology that won World War 2 sired the expansion of air travel. As

Below: *A DH18 at Croydon in 1922, showing that the forward view from the pilot's open cockpit behind the enclosed passenger cabin was pretty restricted.*

larger and higher performance airliners than the DC-3 arrived in service, air travel — previously confined to the wealthy or the daring — came within reach of the average mortal. More and more flights, offering greater speeds and reliability, became available to more destinations: the main down sides were that more aircraft aloft offered more likelihood of collisions, and increased passenger-carrying capability meant more casualties in such an eventuality.

In 1952 the de Havilland Comet introduced the first jet passenger service, but the most popular airliners of that time were the Lockheed Super Constellation and Douglas DC-7, which represented the acme of piston-engined transport development. On the morning of Saturday, 30 June 1956, both types departed from Los Angeles International airport. The first to take off was Super Constellation N6902C belonging to Trans World Airlines. TWA Flight 2, with 64 passengers and six crew on board, was bound for Washington, DC, stopping off at Kansas City *en route*. It was followed into the air three minutes later by United Air Lines DC-7 N6324C. United Flight 718 was scheduled to carry 53 passengers and five crew first to Chicago and then to Newark, New Jersey. Both airline crews were cleared to fly airways, the Super Connie at a true air speed (TAS) of about 310mph and the DC-7 some 20mph higher. While crossing the Mojave Desert, TWA2 asked for permission to climb to 21,000ft to get out of cloud. As this level had already been allocated to United 718, the request was denied by Los Angeles Centre. The TWA crew was, however, cleared to fly 1,000ft 'on top of' the general cloud layer to stay VMC. Unfortunately, this turned out to be 21,000ft. Height separation having been eroded, only time difference remained, but as the DC-7 was flying faster, it was inexorably overhauling TWA2.

Pressurised cabins were but one of many postwar improvements, and it was common practice at the time to allow crews blessed with a high altitude capability to take advantage of the most favourable winds and weather. Once clear of airways therefore, both TWA2 and United 718 took up direct courses northeastward to their respective interim destinations. N6902C flew to the north of the DC-7 but their paths were destined to cross over Grand Canyon

Above:
A TWA Lockheed 1049 Super Constellation similar to that involved in the midair collision over the Grand Canyon.

National Park, some 70 miles north of Flagstaff, Arizona.

In their final regular position updates to Salt Lake Centre, both flight crews estimated that they would reach the Painted Desert at the identical time of 11.31hrs. There would be no further communication with either airliner until the following garbled message was sent by the DC-7's first officer some 30min later: 'Salt Lake ... seven eighteen ... we are going in!' The following day, the wreckage of the two aircraft was found scattered and burned in the Grand Canyon near the confluence of the Colorado and Little Colorado rivers. As everyone on board both aircraft perished, this macabre milestone marked the first commercial aviation disaster to claim over 100 lives.

Subsequent investigation confirmed that 128 people died because a midair collision had taken place. It occurred around 21,000ft, and at the time of impact the two aircraft were angled approximately 25° relative to each other with the DC-7 slightly above. The DC-7's left aileron tip appeared to have caught the Super Constellation's centre tail fin. Straight afterwards, the DC-7's lower left wing surface struck the upper after-fuselage section, during which time the DC-7's No 1 propeller inflicted a series of cuts in N6902C's aft baggage compartment, causing explosive cabin decompression. It all took less than half a second before TWA2 pitched down, turned on its back and plummeted inverted into the northeast slope of Temple Butte. The United DC-7, having lost about 20ft of its left outer wing in the collision, fell less steeply in a curving path. It eventually crashed near the top of Chuar Butte, approximately a mile northeast of the TWA impact site.

The public hearing into the accident considered the wisdom of allowing both airliners to reach the Painted Desert at the same altitude and at the same time. The air traffic controller involved explained that as the Painted Desert reference point was actually an imaginary line some 175 miles long running from a navigation aid in Winslow, Arizona, to another in southern Utah, he had no way of knowing exactly where the aircraft would cross that line. He further stated that since both aircraft were then in uncontrolled airspace, their crews would have been operating under the 'see and be seen' remit inherent in VFR procedures. Yet windows were kept to a minimum in airliners of that generation to maintain the structural integrity of pressurised cabins. Notwithstanding cockpit visibility limitations, the crew of the Super Constellation would have needed eyes in the backs of their heads to have seen the DC-7 coming up from behind and at a shallow crossing angle. Perhaps both crews were preoccupied with trying to give their passengers a more scenic view of the splendid canyon below.

A number of improvements came out of this tragedy, including increased funding for air navigational facilities and modernisation of the US air traffic control system. But technology could only do so much, as was shown on 16 December 1960 when the same two carriers vied once more for the same piece of airspace.

United Flight 826, a DC-8 — Douglas's first jet transport designed to meet the challenge of the Comet 4 and Boeing 707 — had left Chicago for New York's Idlewild International airport carrying 77 passengers and seven crew. TWA Flight 266 was also on domestic duty from Columbus, Ohio, to the other major New York airport, LaGuardia. Coincidentally, TWA266 was an identical Super Constellation (N6907C) to that lost over the Grand Canyon, only this time it was carrying 39 passengers and a crew of five.

The weather over the 'Big Apple' on this Friday morning was overcast with a ceiling of about 5,000ft, accompanied by fog and snow or sleet: these were not natural VFR operating conditions. As United 826 passed Allentown, Pennsylvania, just prior to descent from 25,000ft, it was given a shortened routeing to the Preston Intersection, defined by two radials from two VOR stations where the DC-8 (N8013U) was expected to enter a 'racetrack' holding pattern to await final approach and landing clearance.

Unbeknown to New York Centre, one of the DC-8's twin VOR receivers was inoperative. Consequently, the crew would have had to establish the holding fix by tuning the single receiver from one station to the other, a time-consuming process not eased by the shortened routeing. The last message from the jet was 'Approaching Preston at five thousand'. Centre advised the crew that radar service was terminating and that they should contact Idlewild Approach Control.

Simultaneously, the Super Constellation was descending for its own landing. It was being vectored in on radar by LaGuardia Approach Control, culminating in an instruction to turn on to a heading of 130°. After noticing an unidentified blip on his screen, the controller advised TWA266 that there appeared to be jet traffic '...off your right, now three o'clock at one mile, northeast bound'. The two contacts then merged.

The airliners collided at 10.33hrs while in cloud at 5,000ft almost directly over Miller Army Air Field on Staten Island. Flying up Airway Victor 123 from the southwest at that stage in its flight, the DC-8 crew should have let the speed decay so that they could transition straight into the final approach. But in maintaining an excessive 350mph, they allowed the DC-8 to overhaul the Super Constellation. The slower piston-engined airliner was turning left towards LaGuardia when it was struck from behind and on its right-hand side. United 826 collided with the top of TWA266's fuselage at an angle of about 110°. Cabin insulation and human remains were later found to have been sucked into the DC-8's right outboard engine.

This No 4 engine and associated section of DC-8 outboard starboard wing went down with the Super Constellation as it broke into three main sections before hitting Miller Field. The rest of the DC-8 continued in a northeasterly direction for another 8.5 miles before plunging into the Park Slope section of Brooklyn and bursting into flames. One passenger from the jet, 11-year-old Stephen Baltz riding in the rear of the cabin, was thrown into a snowbank still alive, but he was to die the next day of severe burns and other injuries. Three people were pulled alive from the wreckage of the Super Constellation, but they too died shortly afterwards. The loss of 128 people from both aircraft was identical to the death toll over the Grand Canyon, but the final tally was 134 because six died on the ground in Brooklyn.

This accident occurred because the DC-8 was in the wrong place. Instead of being over the Preston Intersection, United 826 was nearly 10 miles or around two minutes' flying time further on. Coincidentally, this equalled the length of the short-cut passed while the DC-8 was descending from 25,000ft, and accident investigators concluded that the pilots did not take note of the change in time and distance associated with the new clearance. To compound the misunderstanding, there were indications that, in lieu of the second VOR unit, the crew had used its automatic direction finding (ADF) equipment to take cross-bearings. Had No 1 ADF been tuned to the Scotland beacon, the pictorial display would have resembled that expected from a serviceable VOR. Misinterpretation, compounded by the rapid mental agility demanded by mating VOR with ADF calculations, most probably led the crew astray. What is certain is that when the United crew reported that they were approaching Preston, they were already well beyond it.

Unlike the Grand Canyon tragedy, these two aircraft were operating under instrument flight rules (IFR), but being under the watchful eye of a ground radar did not save them. As New York Centre did not adequately monitor the DC-8 and the controller failed to notice when United 826 passed

Opposite above: *A United Air Lines Douglas DC-7 identical to this was the second airliner involved over the Grand Canyon.*

Opposite below: *A United DC-8, similar to N8013U which collided with a TWA Super Constellation over New York City on 16 December 1960. N8013U was the first US commercial jet to crash fatally during passenger service.*

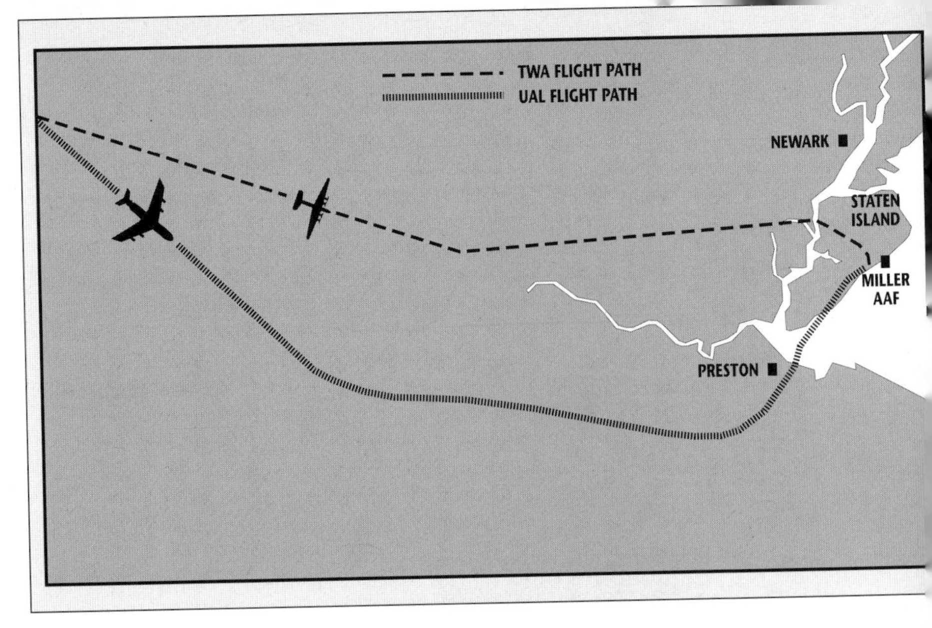

Above:
The flightpaths of the United and TWA airliners which led to them colliding over Staten Island.

Left:
The Park Slope area of Brooklyn after the United DC-8 fell to earth.

the Preston Intersection, this accident triggered another revitalisation of the US air traffic control system. Most significant was the development of the principle of 'positive control' whereby radar control would be exerted from the ground over all aircraft flying above FL240. Positive control would also extend down to 8,000ft on high-density airways and to all airline operations involving jets. A new speed limit was imposed below 10,000ft and within 30nm of a destination airport, and stricter guidelines were imposed on the transference of flights from one control facility to another.

Additionally, two devices were made mandatory on all aircraft weighing over 12,500lb before they could operate in US airspace. The first was the transponder, which enabled any aircraft to impose an individual designator on its 'blip' on a ground controller's screen. The second was distance-measuring equipment (DME) to enable crews to determine their exact position by calculating the mileage between navigational beacons. Subsequent refinements included three-dimensional radar. Since then, this positive control system designed to provide separation of passenger-carrying flights has worked exceedingly well. Not since TWA266 was scattered over Miller Field and United 826 plunged into Brooklyn has a catastrophic midair collision occurred between two large commercial aircraft over the United States.

Would that the skies were only occupied by large commercial aircraft, but aeronautical life is not that simple. On 6 June 1971, McDonnell Douglas DC-9 N9345 departed Los Angeles International airport bound for Salt Lake City. Wearing the livery of Hughes Air West and designated Flight 706, the DC-9 was ostensibly operating under the protection of IFR and positive control. Yet less than 10min after take-off, and while climbing out to the northeast, the DC-9 collided with a US Marine Corps F-4B Phantom heading south to El Toro Marine Air Station. Impact occurred around 18.10hrs at 15,000ft some 20 miles northeast of downtown Los Angeles. The fighter hit almost at right angles, its fin and right wing slicing through the DC-9's lower left cockpit and passenger cabin respectively. Both aircraft plummeted into the San Gabriel Mountains, killing all 49 on board the DC-9 and the Phantom pilot who was unable to jettison his canopy. The fighter's back-seater — the radar intercept officer — ejected to safety to become the sole survivor of the disaster.

The cause of this collision lay fairly and squarely with the failure of both flight crews to see and avoid each other's aircraft. Yet the NTSB recognised that the crews' ability to detect one another, assess the situation and initiate evasive action was minimal. All the classic extenuating factors were here. First, there was the high closing speed of around 750mph. Second, both aircraft were on an essentially constant bearing relative to one another, and would thus have remained almost stationary from each other's perspective. Third, the Phantom had suffered an oxygen system leak and was being flown below normal cruising altitude, which just happened to be a busy airline altitude. Fourth, the fighter's transponder chose that time to go on the blink: a temperature inversion in the area also meant that the fighter did not show up well enough on ground radar to be noticed by a busy controller otherwise unaware of its presence. Finally, the F-4's radar was in mapping mode rather than search mode, and the fighter pilot opted to fly VFR rather than seek deconfliction assistance from air traffic control. Even after all this, the radar intercept officer did see the airliner after lifting his eyes from his radarscope.

He shouted to his pilot who initiated a left roll, but it was too late. As for the DC-9 crew, it probably never saw the F-4B. Successful look-out and scanning techniques need constant practice, and the airliner crew would not have been the first (or last) to be lulled into that false sense of security which comes from over-reliance on ground controllers to provide traffic separation.

Ground controllers and advanced technology can only achieve so much, as was highlighted by the disaster on 31 August 1986 that had been predicted and feared for many years — a collision between a big commercial airliner and one of the many light aircraft that flitted over and around the Los Angeles area.

Flight 498 was an Aeromexico DC-9 (XA-JED) scheduled to arrive from Tijuana at Los Angeles International under IFR and positive radar control. Meanwhile, a single-engined Piper Archer (N4891F), carrying a married couple and their daughter in three of its four seats, was heading for Big Bear in the San Bernardino Mountains of southern California. The Archer was operating under VFR and should, in consequence, have remained outside the Los Angeles Terminal Control Area (TCA). But it did not. Heading east and only eight minutes after getting airborne, the Archer strayed into restricted airspace and into a crossing path with Flight 498 which was then in the descent. Despite clear weather and a visibility of around 15 miles, the two aircraft collided at right angles around 6,500ft. The horizontal stabiliser of the DC-9 sliced into the upper cockpit of the Archer. The airliner rolled on to its back and plunged steeply nose-first into a residential area some 20 miles from the airport: the light aircraft crashed into an open schoolyard, disintegrating in a ball of fire on impact. All 58 passengers and six crew on board the DC-9 were killed, the three occupants of the Archer were decapitated in the collision, and a further 15 people died on the ground.

At first this accident might appear to have resulted from a lackadaisical approach to restricted airspace. But the Archer pilot was found to have been professional and methodical in his approach to flying, with an awareness of the Los Angeles TCA and the regulations pertaining to its use and avoidance. Moreover, a TCA chart was found in the cockpit wreckage. Given the facts, plus evidence to show that the Archer had remained under control up to the collision, the NTSB concluded that the 53-year-old pilot flew into the control area inadvertently, probably after misidentifying his navigational checkpoints.

Once again, a factor in the accident was the limitation of the 'see and be seen' principle in ensuring separation. Tests demonstrated that each aircraft should have been visible to the other's crew in sufficient time to avoid the collision, especially in the case of the private pilot: however, there was no evidence of any prior sighting or evasive manoeuvring. Ironically, the Archer carried a functioning transponder, and a tape-recording of the radar ground display did show the echo of N4891F. Unfortunately, the ground controller did not see it at the time, not least because he had been distracted by a second light aeroplane that had inadvertently penetrated controlled airspace. Be it in the air or on the ground, 'seeing' is one thing; allowing the brain enough time to register and react to what has been seen is another.

And as aircraft cockpits eschew dials and gauges, becoming more glass-screened and high tech such that they resemble a cross between a video arcade and a television shop, so the temptation grows for aircrew to remain heads-down in the cockpit rather than out scanning for possible conflictions. The great airliner sedately going about its business along the tightly monitored airway is no longer a problem. It is down where military jets practise flying fast beneath radar cover, alongside the smaller fry of civilian aviation trying to make a living, that the last bastion of 'see and be seen' is still to be found.

On Wednesday, 23 June 1993, Agusta-Bell 206B JetRanger III helicopter G-BHYW, flown by a highly experienced freelance pilot, set off across Lancashire with a pipeline superintendent on board as part of a routine series of pipeline inspections carried out on behalf of Shell Chemicals UK Ltd. After refuelling at Blackpool airport, the helicopter took off at 10.19hrs on the next phase of its inspection. The JetRanger followed the pipeline which ran parallel to the M6 motorway, maintaining a flight information service with the British Aerospace airfield at Warton until radio contact was lost due to terrain screening. At 10.52hrs, the helicopter arrived in the vicinity of Farleton Knott near Kendal, Cumbria, where three engineering sub-contractors were working on the pipeline. The sub-contractors reported that the weather at the time was 'clear, sunny and fine with good visibility, enough to see the Lake District hills'. The pilot flew an orbit over the trio; the ground workers stopped working, waved and were able to see clearly the face of the observer waving back. On completing the orbit, the pilot rolled out straight and level at about 300-400ft above ground level (AGL) on a northerly heading to follow the pipeline route. Moments later the JetRanger was struck by a Tornado GR1 bomber.

The Tornado GR1 is the RAF's primary and furthest-reaching strike/attack aircraft. Designed to penetrate the massed defences of the former Warsaw Pact, the Tornado GR1 could achieve its aim only by going in at low level. Successful low level operations are not something that can be done at the drop of a hat — they have to be practised regularly to achieve and sustain the confidence necessary to underpin proficiency. Unfortunately, RAF squadrons based in Germany had the training ground cut out from under them in November 1990 when the German government unilaterally imposed a 1,000ft minimum limit on all military flying over Germany. The only alternative was to train over the UK, and so it came about that Tornado ZG754 departed RAF Brüggen at 09.44hrs on 23 June to fly as No 2 of a pair of Tornados on a sortie involving various bombing profile manoeuvres using an air-to-ground weapons range on the east coast of the UK. With the range work complete, the crews planned to carry out a low level attack transit before landing at RAF Leuchars, Fife, to refuel.

Navigational warnings were checked at Brüggen during initial planning on 22 June and again on the morning of

Opposite: *The scene of devastation after the Aeromexico DC-9 hit a Los Angeles suburb.*

the sortie. ZG754 took off late following a minor technical problem, but rejoined the leader on the Cowden weapons range. Thereafter, the two aircraft continued in a tactical formation — lateral separation of about 4,000yd — on the low level portion of the route. No 2 flew on the leader's right, the pilot reporting that visibility at low level was excellent.

The first significant turn *en route* was to be right, around the town of Kendal. Three minutes before the turning point, the Tornado crews were following a route parallel to the A65 trunk road displaced approximately 3nm to the south. As they approached the high ground of Farleton Knott, the pilot of ZG754 decided to fly tactically: he followed the A65 valley to the north of Farleton Knott while his leader flew to the south. Now surrounded by high ground, the pilot of the No 2 aircraft lost sight of his leader and concentrated on ensuring that the area ahead was clear. As he reached the end of the valley and knowing that they were soon to initiate a turn, he looked left for the other Tornado. The navigator in the rear seat, his forward vision restricted by aircraft equipment, concentrated on the area to his right where the chart showed an active hang-glider site. As he glanced to his left, the pilot heard a loud bang.

The Tornado's pitot probe struck the bullet fairing aft of the JetRanger's tail rotor almost at right angles. Following the collision, which severed the helicopter's tail boom just aft of the horizontal stabiliser, witnesses saw light debris fall from the helicopter as it entered a series of three descending spirals to the right before stabilising at about 150ft. It then fell vertically to the ground in an upright attitude.

Below: *A Tornado GR1 bomber in its element at low level.*

Opposite: *The impact of a JetRanger tail rotor on a Tornado radome.*

The tail rotor and gearbox assembly penetrated the Tornado's radome and was ingested by its starboard engine. Fearing a major bird strike, the Tornado crew diverted to Warton airfield where they landed without further incident. A policeman, who witnessed the accident, alerted the rescue services which arrived quickly. The helicopter crew, however, was pronounced dead at the scene.

The collision occurred at a height just below 400ft AGL, with the Tornado covering the ground at 440kt on a heading of 304° and the JetRanger at an estimated 60kt heading 036°. Numerous eyewitnesses saw the accident but few saw the Tornado before it hit the JetRanger. None noticed any sudden change in attitude or flightpath to suggest any attempt to avoid a collision. Neither crew of either Tornado saw the helicopter before or after the accident.

After such accidents, two investigations are carried out in parallel — one by an RAF Board of Inquiry and the other by the AAIB. Both came to the same conclusion. The collision occurred in uncontrolled airspace at low level and in excellent visibility. There was no evidence to suggest that medical incapacitation or any technical problem contributed to the accident. The RAF found that the Tornado sortie had been planned and briefed correctly, while the AAIB found that the helicopter pilot had been properly checked, he had flown pipeline inspection flights on seven previous occasions and he was assessed as competent on all tasks including pipeline patrols.

In other words, both aircraft and crews had every right to be where they were doing what they did. The experience of the JetRanger pilot and his familiarity with the route would have made him aware of possible conflictions with low-flying military aircraft. He was occupying the right-hand seat and there was no evidence that he was not looking out. On the other hand, many tasks in an all-weather bomber such as the Tornado require crews to look inside the cockpit. That said, evidence showed that the workload approaching Kendal was light and that, for at least 15sec prior to impact, both crew members were devoting their attention to look-out. The trouble lay with the difficulties of seeing and being seen. Modelling techniques showed that the JetRanger chugging along at 60kt would have had relatively little motion relative to the Tornado crew and that the converging aircraft would have appeared extremely small to each crew until at a late stage. Furthermore, in looking left for his leader, the Tornado pilot's head would also have moved left and consequently the helicopter may have been completely obscured behind the windscreen arch. Finally, the JetRanger would not have subtended an angle greater than the thickness of the Head-Up Display symbology until seven seconds before impact.

Eventually both military and civil investigations had to conclude that, despite the good weather with excellent visibility pertaining at the time, the cause of the accident lay with the failure of the

Tornado crew and the pilot of the JetRanger to see each other's aircraft in time to take avoiding action. The Tornado was fortunate in being able to land at Warton despite having sustained considerable damage. The JetRanger stood little chance of surviving the loss of its tail rotor assembly and stabiliser.

The Tornado and JetRanger collided because of a failure in the 'see and avoid' principle, yet this really meant that the true failure lay with the limitation of the human eye. Fitting high intensity strobe lighting or snazzy paint schemes could prove beneficial in some weather conditions, but like the fancy claims made for collision warning systems and electronic detectors, they could only deliver so much. Physiological limitations dictate that, notwithstanding all the safety improvements conferred by technological advances around airlanes at altitude, down among the weeds the same perils await Tornado and JetRanger that faced the DH18 and Farman Goliath. And just as in 1922, maybe the only sure way to be safe is through the strict enforcement of one-way corridors.

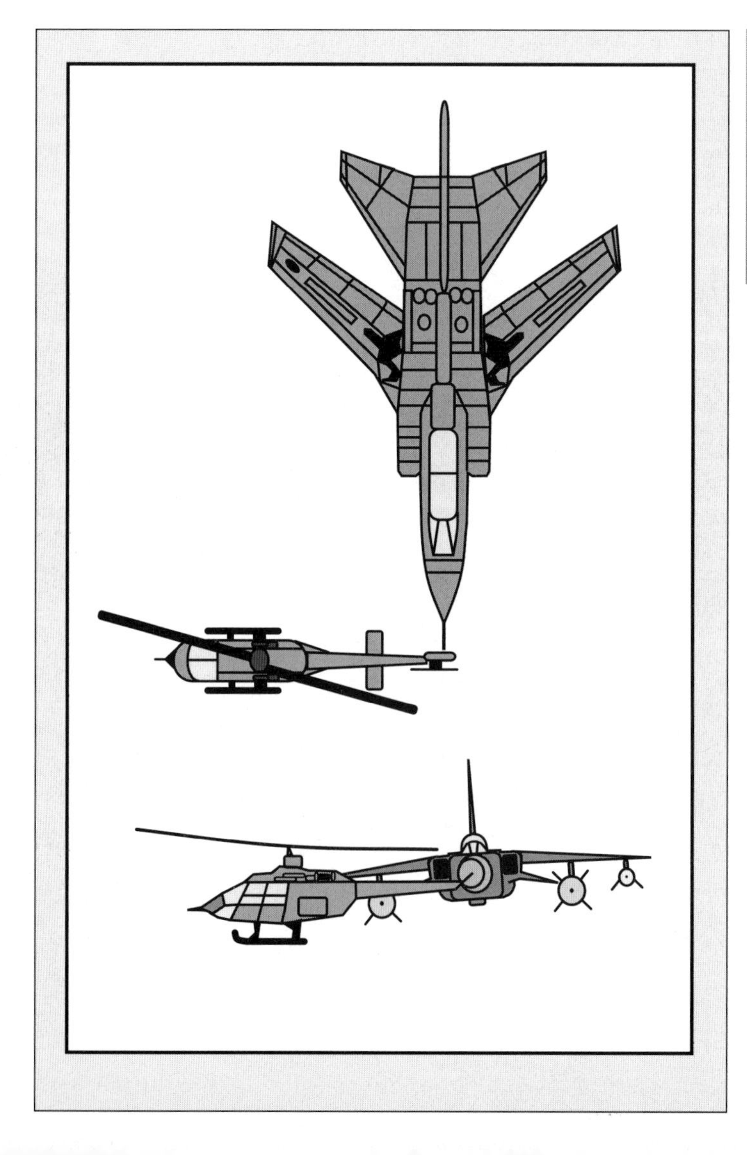

Left: *Diagrammatic representation of how the Tornado's pitot probe struck the fairing aft of the JetRanger's tail rotor almost at right angles.*

4 Design Faults

British military flying came of age in 1912 when the Royal Flying Corps was established around the Naval Wing, the Military Wing and the Central Flying School. Official resolve to procure the best aircraft for them justified the holding of a Military Aeroplane Competition at Larkhill in August 1912, which stimulated a whole range of disparate aircraft sizes and configurations. But soon after the ending of the trials, the Military Wing lost four of its officers in two crashes during the Army's autumn manoeuvres. On Friday, 6 September Capt Patrick Hamilton, with Lt Wyness-Stuart as observer, died when their Deperdussin broke up in the air and crashed near Hitchin. Four days later, Lts Edward Hotchkiss and C. A. Bettington were flying a Bristol-Coanda from Larkhill to Cambridge at 2,000ft when their aircraft began to descend over Port Meadow, Oxford. The descent became a steep dive; at 200ft, fabric tore off the starboard wing and the aircraft plummeted to the ground, killing both officers.

Below: *Wreck of the Bristol-Coanda monoplane after it crashed in Oxford on 10 September 1912.*

These two aircraft accidents had little in common other than that they befell monoplanes, but monoplanes had been having a bad press of late. The first British military heavier-than-air flying fatality befell Lt R. A. Cammell on 17 September 1911. He died when, at a height of around 90ft over Hendon, his aircraft tilted to one side and fell with a crash. His untimely death occurred because he was unfamiliar with his machine's controls: it is believed that he forgot to work the forward elevator. But he was flying a Valkyrie monoplane at the time, and people remembered that long after they had forgotten the cause of the accident.

Five other pilots flying monoplanes perished during the summer of 1912, and even though the accidents had nothing to do with monoplanes *per se*, higher authority felt that something had to be seen to be done, not least because the accidents wiped out a fair proportion of the best Military Wing aircrew talent of the time. The French promptly grounded all monoplanes until they had been structurally modified to a standard laid down by M Louis Bleriot, but the British War Office came up with no such measured response; it simply issued an instruction banning the use of monoplanes by pilots of the Military Wing. The 'monoplane ban' remained in place for six months, resulting in the abandonment of several monoplane types without any structural testing or analysis. But the impact was more far-reaching than that. Among those who signed off the Army Monoplane Report dated 3 December 1912 were several who would become movers and shakers in the wartime Flying Corps. Their prejudice against monoplanes would last longer than their formal ban, and would result in an unhealthy dependency by the Military Wing on biplanes. Such a knee-jerk reaction to flying accidents would be repeated over and over again in subsequent years, and it underscores the point that the best accident investigators should be given the time and facilities to come up with the right cause of any accident, not the one that is the most politically expedient.

Having said that, design faults have plagued air travellers. When the Canadian Government decided to build some four-engined aircraft immediately after World War 2, it obtained the rights to the Douglas DC-4 which was then pressurised and fitted with four supercharged Rolls-Royce Merlin engines. BOAC ordered 22 of these C-4 variants which were christened Argonauts. Argonauts did sterling service on BOAC's African and Far East routes for 11 years before they were sold off. So it came about that Argonaut G-ALHG, now belonging to British Midland Airways, found itself bringing a full load of holidaymakers back from Palma, Majorca, into Manchester Ringway on 4 June 1967. All went well until the airliner was established for an ILS approach to Runway 24. Suddenly, both starboard engines started to malfunction. Capt Harry Marlow transmitted that he was initiating an overshoot because he was having '...a little bit of trouble with the rpm'. The Argonaut completed a 360° orbit to the right, during which time it broke through the low overcast cloud. G-ALHG was back on the ILS localiser with undercarriage still retracted and flap set at 10° when it crashed in the centre of Stockport, some five miles short of the runway threshold. The airliner hit the ground in a nearly-level attitude but its port wing was torn off as it struck a three-storey building.

Initially there were only small, scattered fires, but about 10min after the crash an explosion rent the starboard wing and the blaze spread rapidly to the fuselage. The flames beat back rescuers after 10 passengers had been saved, all of whom were injured, as was the stewardess who was thrown clear through a tear in the cabin. Capt Marlow got out alive though he suffered from retrogressive amnesia and could not remember anything about the flight. The other 69 passengers and three crew members perished. Although many of them survived the impact, they were trapped, and died from suffocation in the inferno. It was the worst casualty list from a fire in a British airliner until 55 died in a Boeing 737 fire on 22 August 1985, coincidentally also at Manchester.

The trigger for the accident was the loss of power on both starboard engines. This should not have posed too great a problem, especially as the Argonaut was lightly loaded at that stage in its flight. But accurate asymmetric flight proved to be so difficult that the flying pilot was unable to maintain height, especially as one of the propellers was left to windmill. But the cause of the malfunction was not mechanical: it lay in the design of the fuel cocks and the location on the flightdeck of their actuating levers. So awkward was their placement on the console that a pilot could relatively easily put one or more of the levers in the wrong position, and only a slight error in the positioning of a lever could create a situation whereby a cross-feed valve would be 'cracked', or improperly closed, allowing for the inadvertent transfer of fuel from one tank to another. This could catch crews out if fuel was transferred from a tank that they remained convinced was sufficiently filled.

The upshot was that G-ALHG's No 4 engine suddenly wound down on the approach due to fuel starvation. No 3 may also have failed for the same reason or, perhaps, because the pilot shut it down after misidentifying the starboard inboard for its malfunctioning neighbour. It was then possible that the pilot got in a bit of a tangle with feathering and unfeathering, but power was not restored to No 3 engine in time to prevent the crash. That said, although the outcome proved disastrous, Capt Marlow was not faulted for executing a go-around under the circumstances. It is easy to say that he should have thrown the Argonaut on the ground as soon as the engines started playing up, especially as he was on final approach at the time, but it is perfectly understandable why he did as he did. There were no indications that he was in a life-threatening situation where time was of the essence, and as he was in solid cloud down to 300ft, the priority was to keep calm and control the aeroplane. When it all went wrong, he showed skill in setting G-ALHG down in the only open space available to him. And it could have been much worse — immediately beyond the crash site were tall blocks of flats, the town hall and police station, and Stockport Infirmary.

The crew's problems were compounded by the poor asymmetric handling characteristics of the Argonaut. Accident investigators noted that the lack of instrumentation to indicate engine failure and the poor design of the fuel system controls would have prevented the C-4 from being certified under 1967 standards. The obvious question to ask is why, after so many years of Argonaut operation, the aviation community only discovered this design fault the hard way so late in the airliner's life? But in truth, other DC-4 operators had experienced cases of inadvertent fuel transfer, but these had never been passed on to the appropriate British Government or airline authorities. Nor had users such as British Midland Airways been adequately alerted to the potential hazard by the manufacturer. British Midland's engineers and pilots did not even realise that such an inadvertent transfer of

Right: *Flightpath showing Argonaut G-ALHG's final minutes.*

Opposite above: *A Canadair C-4 Argonaut, similar to that which crashed at Stockport on 4 June 1967, in an earlier BOAC livery.* Quadrant Picture Library

Opposite below: *Remains of the British Midland Airways Argonaut in the centre of Stockport. Fortunately the airliner hit an open space, thereby avoiding a higher death toll.*

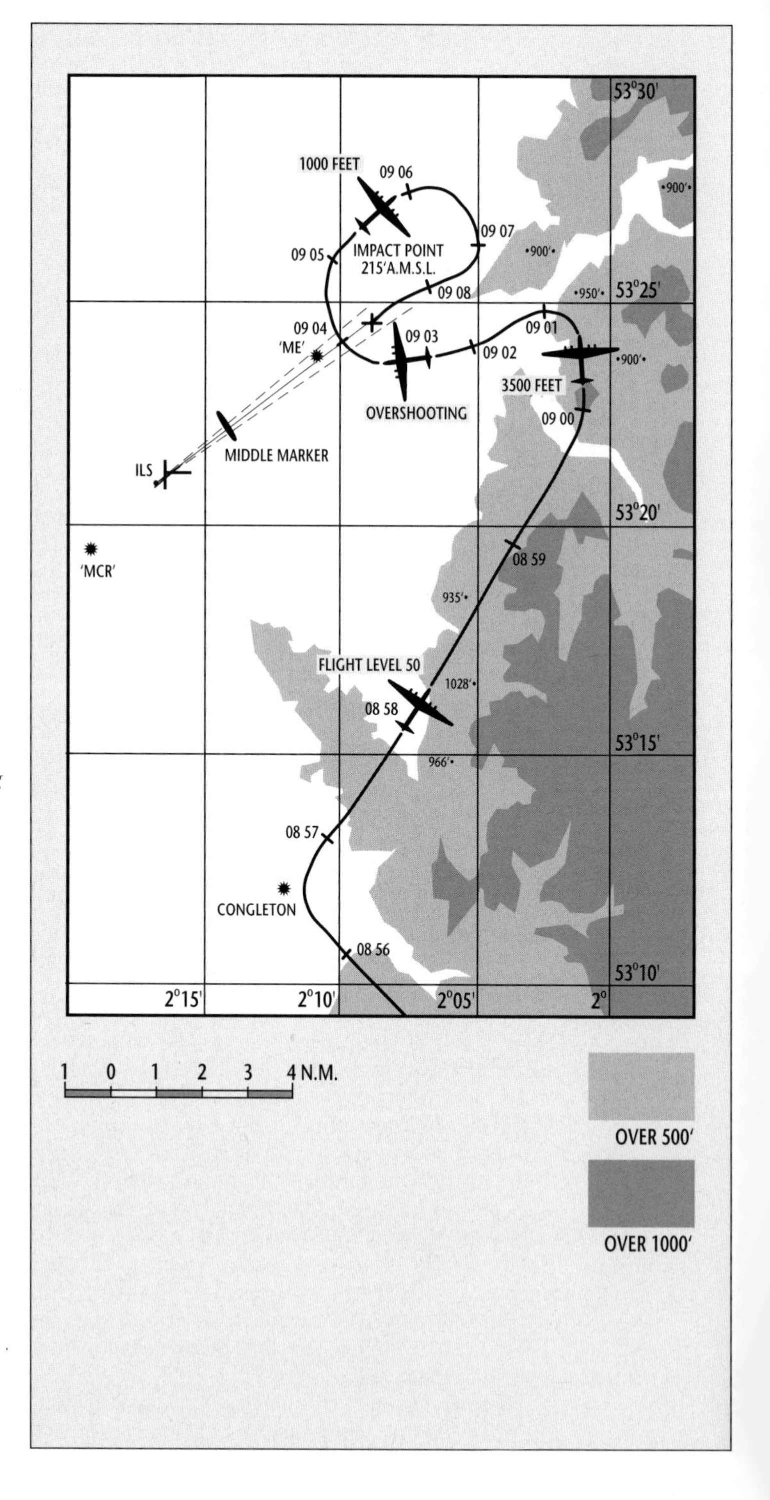

B·O·A·C
G-ALHT

Shaws
FOR
FURNITURE
Shaws
FOR
FIREPLACES

fuel was possible. The great flight safety lesson here is to share information, even if that information might have been obtained in embarrassing circumstances. It takes no great ability to learn from one's own mistakes; the great flight safety aim should be to learn from the mistakes of others.

Some design faults can take a long time to show up, and others can be brought into focus by human frailty. On Sunday, 5 July 1970, Air Canada Flight 621 took off from Montreal bound for Los Angeles via Toronto. Airport weather at Toronto was ideal as the Super 63 version of the DC-8 (CF-TIW) let down over Ontario. It was subsequently clear from the CVR tape that Capt Peter Hamilton and First Officer Donald Rowland omitted the 'spoilers armed' item from their pre-landing checks. They agreed that these lift-spoiling devices would be armed during the flare, allowing them to extend automatically once the wheels made contact with the runway. However, as the DC-8 crossed the threshold at around 60ft up, the co-pilot made an error. When told by Capt Hamilton to arm the ground spoilers, which involved lifting the appropriate lever on the control console, instead he pulled it to the rear, triggering their immediate deployment. Capt Hamilton yelled, 'No, no, no!' and the first officer, realising his mistake, responded with, 'Sorry, Oh sorry, Pete'.

The flying pilot applied full power to all four engines and rotated the DC-8's nose upwards to counter the high rate of sink that had developed. But the spoilers won out in the short height available and Capt Hamilton's actions were unable to prevent CF-TIW from making an exceptionally hard landing. Such were the downward forces generated that the starboard outer engine pod and pylon assembly separated from the aircraft and the bottom of the No 4 alternate fuel tank was punctured. Escaping fuel then ignited on contact with electrical wiring that severed when the Pratt & Whitney turbofan broke off. It was one of those days that justified all the mandatory instructions to passengers and crew alike to 'fasten your seat belts for landing'.

The DC-8 had only impacted with the ground for a fraction of a second before it climbed away. Capt Hamilton

Below: *An Air Canada DC-8 Super 63, identical to the aircraft which crashed on 5 July 1970.*

radioed his intention to fly another circuit and land again on the same runway. The airliner's undercarriage, which had not been badly damaged, was raised for the go-around and its flaps set to 20°. The spoilers were fully retracted.

But some 3min after the hard landing, Flight 621 was rocked by three explosions. The second of these ripped off the No 3 engine pod and pylon, while the third blew away a large segment of the starboard outer wing. Capt Hamilton was unable to maintain control thereafter and the DC-8 plunged from 3,000ft into a field near Malton some five miles north of the airport. CF-TIW hit the ground at a speed in excess of 250mph, so it was not surprising that the airliner disintegrated in a ball of fire on impact, killing all 100 passengers and nine crew on board.

The accident investigative report did not fault the captain for executing the overshoot procedure. The flight crew probably never realised the extent of the damage sustained by the DC-8's glancing blow with the runway, otherwise Capt Hamilton may well have continued the landing after the first officer's error which would probably have been classed as an expensive incident with no casualties.

Air Canada procedures dictated that the spoiler system be armed for automatic deployment at an altitude of 1,000ft, so the pilots were not adhering to company regulations. The flight crew had its method of activating the spoilers but, interestingly, on the day of the accident the pilots did not use the technique they had adopted when flying together previously, which was to extend the spoilers manually after touchdown. The moral here is that, if you are going to refine company rules for whatever reason, you should be doubly alert when it comes to making selections that are out of the ordinary.

But the subsequent Board of Inquiry blamed the disaster as much on faulty design as on human error. It argued that while a single actuating mechanism might be acceptable to perform different secondary tasks such as heating or ventilation, this arrangement was completely unacceptable for something as crucial as lift spoilers. The Board believed that, at the very least, the activating lever should have been fitted with some guard or gate. The wisest precaution would have been a 'weight-on' microswitch that only closed to allow current to pass to the actuator once it sensed the aircraft had made contact with the ground.

Furthermore, investigators found that the instruction manuals provided by the manufacturer were so misleading and inaccurate that they led readers to believe that there was a mechanism built-in to prevent the spoilers from in-flight extension. Consequently, Air Canada training staff did not realise that such inadvertent deployment as befell Flight 621 was possible. This ensured that the airline's pilots were never made aware of the potential threat to flight safety. In passing, the Board also criticised the design of the pod/pylon structure and what it described as the failure of the manufacturer to ensure the integrity of the fuel and electrical systems built into the DC-8.

As a result of the loss of Air Canada Flight 621, the FAA issued an Airworthiness Directive requiring placard warnings against in-flight operation of ground spoilers by DC-8 operators. It was only after a non-fatal accident some three years later that the FAA issued another directive requiring that all DC-8s be fitted with spoiler-locking mechanisms to prevent such an occurrence. It is pleasing to know that accident investigations can bring

about such remedial action; it is a shame that such an obvious safety device, which could have been incorporated at the initial design stage for minimal cost, only became mandatory some 12 years after the DC-8 first flew.

But some design failings do take positive human contrivance to come to light, especially if an aircraft is used for purposes other than those its designers intended. Early on the morning of Friday, 27 August 1993, an Antonov An-28 twin turboprop (HA-LAJ) operated by the Hungarian Aeronautical Association left Budapest with two Russian pilots on board. They refuelled at Maastricht before continuing on to Weston-on-the-Green, a RAF grass strip used for parachute training. The An-28 was on hire to the RAF Sport Parachute Association. With its Russian crew well rested, the aircraft completed a series of 12 flights on Saturday, 28 August, each of approximately 15min duration, in support of freefall parachuting over the airfield.

On the thirteenth flight, HA-LAJ departed Runway 36 with 19 parachutists and crew on board. As the An-28's flaps fully retracted at approximately 500ft, all on board were alarmed to note both engines simultaneously suffering total power loss. The aircraft's speed then decayed rapidly, leading the 40-year-old commander to suspect that the automatic outboard wing spoilers had also deployed. Loss of power combined with an increase in aerodynamic drag from the spoilers left a forced landing as the only option open. The commander initiated a steep descent to maintain speed and turned through 90° to the right to position himself for a forced landing on the only large area available — a field of corn stubble adjacent

Above: *The shattered remains of Air Canada Flight 621 scattered over the countryside around Malton, Ontario.*

to the airfield. The high-winged An-28 with its fixed landing gear impacted heavily in a slightly nose-up attitude, banked slightly to the right, at an estimated speed of 92kt. Even though the harvest had been taken in, the aircraft landed heavily and was damaged beyond repair. Fortunately, the passengers and crew were able to vacate the aircraft without injury.

The commander and co-pilot had previously operated the An-28 in a parachute dropping role in Hungary; they were properly qualified and they fulfilled their duties well in good weather conditions. Their aircraft had been refuelled several times during the day and it had adequate fuel reserves on board at the time of the accident. According to the commander, the take-off and climb on the fateful thirteenth trip were flown in accordance with recognised procedures and correct speeds. At the flap retraction height of 500ft, the co-pilot selected flap 'in' by three rapid blips of the selector switch. It was while he was doing this that both engines wound down. Crew, occupants and ground witnesses agreed that both engines lost power at the same moment, and both propellers were in the feathered position at the time of impact.

Damage to the airframe was consistent with a high rate of descent at touchdown, which was understandable given the total power loss plus increased drag from the spoilers. Since a spoiler could only be opened after propeller feathering and engine shutdown, and as both spoilers were open during the final descent, both left and right propeller feathering and engine shutdown systems must have operated inadvertently. But why did the engines decide then and there to fail in unison?

In the light of the evidence and the pilot's recollections after his return home, the Russian Department of Air Transport studied the wiring diagrams and system schematics of the An-28 and consulted the Antonov Design Bureau. They came up with a hypothesis which centred on the fact that signals from the electrical flap selection switch were routed via a terminal block, at which point the circuit became common with the circuits of the two fuel shut-off/autofeather/asymmetric spoiler-deployment systems (one such circuit being energised when an engine lost power during operation at a high throttle angle). Completion of these two circuits to earth was achieved by earthing the complete terminal block via two cables and tags secured to the structure by a single screw. The Russians carried out a trial on another An-28 and they found that the presence of a dormant loose earthing connection at a common earth terminal permitted an electric current to flow from the flap signalling system, upon flap retraction selection, to both propeller autofeathering systems. This automatically closed both engine fuel shut-off valves and operated both asymmetric spoiler systems.

On the face of it, a design that allowed a simple flap-up selection to have such a major impact on aircraft safety was not the most adroit in aeronautical history. In Antonov's defence, its An-28 had apparently been designed with dual earthing points for the flap signalling and propeller autofeathering circuits, but during manufacture of the type in Poland, a single earth installation for both systems had been implemented locally as the build standard. This did not matter until the single earthing screw progressively loosened.

It became apparent during the accident investigation that the Hungarian authorities had not notified their Russian counterparts that the An-28 had been transferred to Budapest's books. The debate over registration was not an irrelevance. The Hungarian aircraft flight manual supplement permitted the aircraft to be used for parachute jumping and also gave clearance for flight with the rear clam-shell doors removed. The Ukrainian Antonov Design Bureau on the other hand was emphatic that its aircraft had not been evaluated for flight without the rear doors fitted, and the Bureau considered it to be unsuitable for such operations. It stated that the guide on flight operations for the An-28 specifically prohibited its use for parachute training. The failure to notify the Russians of the registration of the aircraft in Hungary, a simple oversight lost in the disorganisation that followed the demise of the Soviet Union, denied the Russians the opportunity of informing the Hungarian authorities that this type had not been cleared for the role in which they intended to use it. A Hungarian-Russian bureaucratic tangle, combined with the fact that an aircraft designed by the Ukrainian Antonov Design Bureau had been unilaterally modified by Poles, illustrated how easily political intertwinings can mask an aircraft accident waiting to happen.

5 Climb Every Mountain

Every nation has its élite fighting units. The German *Gebirgsjäger*, or mountain troops, were just such an outfit. Their battlecry — 'Hurray the Chamois' — reflected their belief that, like the nimble deer which inhabited the most isolated mountain regions, they could survive and flourish among some of the most difficult and inhospitable terrain on earth. But the chamois needed to be fleet of foot because it had no wings.

The *Gebirgsjäger* were formed in the Tyrol and they saw action mostly in the mountain regions of Norway, France, the Balkans, Crete, Russia and Italy. Typical of their number, with edelweiss flower emblem on uniformed jacket, was Eduard Wolrath Dietl. Dietl was awarded the Knight's Cross in May 1940 when, after leading his troops in a difficult action during the battle for Norway, he found himself dubbed the 'Hero of Narvik'.

For four years after that triumph, Dietl and his *Gebirgsjäger* fought in Finland and Lapland. In June 1944 Generaloberst Dietl visited Hitler's southern mountain retreat at Berchtesgaden high above Salzburg. It was Dietl's intention to return to Norway by air via Graz and Vienna. On 23 June he took off from Graz together with a number of other military personnel — including three generals — in a Junkers Ju52. The Ju52/3m was probably rivalled only by the Douglas DC-3 as the most famous transport aircraft of its time, and this tri-motored beast was sturdy enough to cope with most eventualities apart from flying into mountains.

The pre-flight forecast from the Graz meteorological office predicted cloud cover between 1,500 and 7,500ft. The Ju52 pilot was therefore advised to fly around the Alps over the plains on his way to Vienna. This was sound advice for although Dietl's pilot had many flying hours under his belt, he was not experienced in mountain flying. But the Hero of

Below: *The last goodbye. Gen Dietl waves before boarding his Ju52 on 23 June 1944.*

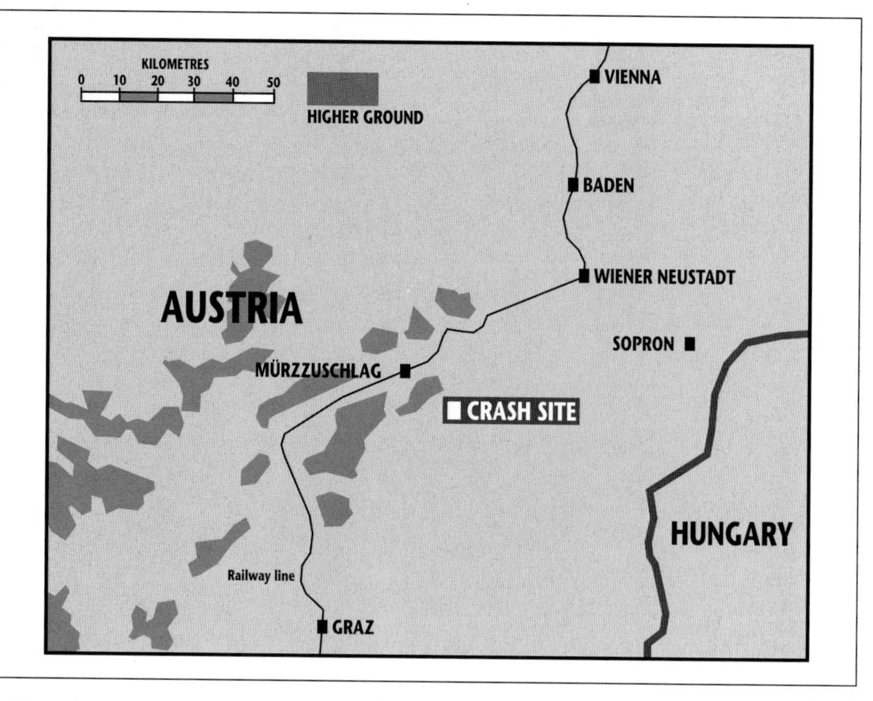

Narvik loved his mountains and, as by far and away the most senior officer on board, he instructed the pilot to fly the direct route over the Alps because it would be a change from the flatness he was used to in Finland. With great confidence he pointed out that the clouds were high and that he knew the ranges well.

Being just an Oberleutnant (captain), the pilot did as he was told. From Graz he flew the Ju52 north down the valley before turning right along the Semmering railway line. Passing over Mürzzuschlag, the ground rose very sharply before him to over 5,000ft where it met the clouds pressing down. If the pilot had thrown it away there and then, and carried on flying down the railway line to Wiener Neustadt, he might have got away with. But he then seems to have turned right, possibly under instructions to go sightseeing. Despite flying lower and lower to stay in visual contact with the ground, the pilot lost the peaks ahead in the mist and cloud. He flew the lumbering Ju52 over one ridge but then entered cloud near the little village of Rettenegg. Going onto instruments, the pilot wheeled 95ft of wing span round to try and go back whence he came, but it was too late. The aircraft hit a group of pine trees at around 120mph before coming to grief on the mountainside at Hartberg am Semmering, killing all nine on board.

Generaloberst Dietl's body was retrieved from the isolated mountainside and returned to his home town of Munich for burial. He was posthumously awarded the Swords to his Knight's Cross, and it was a measure of the esteem in which he was held that Hitler delivered the address at his state funeral — a rare public appearance by

Above:
Map showing Graz, where Dietl took off on his final flight, and his intended destination, Vienna. The crash site near Mürzzuschlag is also shown.

the Führer at that stage in the war. It is always ironic when someone who has survived and prospered through much adventurous action finally loses their life in an accident, but it is doubly ironic if that accident was avoidable. The short flight from Graz to Vienna would have taken barely an hour even by the curving easterly route skirting the Alps. But Dietl, the superb mountaineer, was known as a leader who drove his men hard, and he pressed his pilot, doubtless against the flyer's better judgement, to take the direct route to disaster. It is easy with hindsight to say that Dietl had enough experience to know that cloud-covered mountain valleys were death-traps, and that he should not have ridden roughshod over the professional advice of his pilot who was far from experienced in operating over mountainous terrain. But that is easier to say than do, especially when tunnel vision takes over and you just have to satisfy that urge or meet that deadline. Pressure on pilots, placed there by higher authority or by themselves, should never become so great that it becomes impossible to say 'No'.

Another great leader, ruthlessly determined and known for sparing neither himself nor his subordinates in the pursuit of his objective, was Air Chief Marshal Sir Trafford Leigh-Mallory. Most famous now for his advocacy of the 'Big Wing' concept while in command of No 12 Group during the Battle of Britain, Leigh-Mallory had gone on to bigger and better things such that by 1944 he was given overall command of the Allied Air Forces in southeast Asia at the age of 52. As the war in Europe was drawing to a close, Leigh-Mallory knew that fate was offering him a starring role in the final act of World War 2 — the sinking of the Rising Sun. It must have been with great anticipation that this ambitious man and his wife, Doris, climbed aboard Avro York MW126 on the morning of Tuesday, 14 November 1944.

First designed and flown in 1942, the Avro York employed the same wings, undercarriage and engines as the Lancaster bomber but it had an entirely new fuselage with twice the Lancaster's cubic capacity. Although lacking in creature comforts such as cabin heating, the York had a maximum speed of 298mph at 21,000ft and a range of 2,700 miles. Given that the British aircraft industry concentrated on fighters and bombers during the war, the thin trickle of

Below: *Avro York MW102, the personal aircraft of Lord Louis Mountbatten. A sister aircraft, MW126, was allocated to Air Chief Marshal Leigh-Mallory.*

Yorks prior to 1945 consisted mainly of aircraft for VIPs who had far to go. And Leigh-Mallory had some 5,500 miles to go to his new headquarters in Kandy, Ceylon.

York MW126 was practically new when it was handed over to serve as Leigh-Mallory's personal aircraft. On 2 September 1944, Flt Lt Charles Lancaster DFC and bar, a 32-year-old Cambridge graduate and possessor of an engineering degree plus over 2,000 flying hours and excellent flying assessments, was interviewed by Leigh-Mallory and appointed as his personal pilot. Other crew members simultaneously selected were Flt Lt Peter Chinn, the ACM's personal assistant who doubled as a second pilot, Flt Lt Keith Mooring (navigator), Flt Lt John Casey, RAAF (wireless operator), Flg Off Alfred Enser (flight engineer), two aircraft fitters and Sgt Harold Chandler (steward).

Charles Lancaster, now promoted to squadron leader, was currently on Sunderland flying-boats, so he converted on to the York at RAF Lyneham. He amassed the princely total of 9hr 40min on type, of which four hours were solo, before he brought MW126 to RAF Northolt, west of London, on 3 November to be housed in the York Flight. Leigh-Mallory's York carried standard wireless and navigational equipment but when it arrived at Northolt, the internal fittings and seating were modified to the ACM's personal requirements. The four Merlin engines, their propellers and the undercarriage were brought up to the latest Air Ministry standards and strengthened beams were fitted, all of which 'materially increased the weight of the aircraft' to 43,517lb. The modification work, including that necessary for flying to the Far East, was carried out by men experienced in servicing Yorks, and on 13 November MW126 was flight tested and found serviceable. The fact that the York was fitted with fluid de-icing equipment rather than windscreen wipers excited no comment.

The York was refuelled with 1,860gal of fuel and the previously weighed baggage was loaded. On the evening of 13 November, the flight crew were briefed on the terrain, restricted areas and emergency diversion airfields *en route* to their first staging post at Pomigliano, just east of Naples. The navigator, Flt Lt Mooring, asked for a set of astronomical and navigational tables and confirmed that he required nothing further. Flt Lt Casey was issued with an amended copy of Route Signals Instructions and a map of France and Italy showing all the radio aids to Naples. The Met Officer told the crew that the cloud tops would be around 12,000ft and up to 15,000ft in showers. Any cumulo-nimbus cloud above freezing level was expected to stimulate the formation of heavy, clear ice. When Sqn Ldr Lancaster asked if he would be above the cloud at 17,000ft, he was told, 'Generally, yes.'

At 09.07hrs on 14 November, the York with its distinguished passengers took off from Northolt for the 1,120nm trip to Naples. As it passed 1,000ft, it joined up with 11 orbiting Spitfires which escorted the York to overhead Selsey Bill. The fighter escort then broke away, leaving MW126 to fly steadily on across the English Channel. Overseas Aircraft Control, Gloucester, received the transmission, 'Goodbye Fighter Command. Many thanks escort.' It was the last message heard from MW126.

At around 10.00hrs, the pilot of a Dakota travelling from Lyneham to Naples sighted a York about 1,500yd away on his port side, near the town of Coutances in Normandy, flying on a southerly heading. His reaction on seeing the York was that it was 'probably in the freezing level,' though it appeared to be flying normally with none of its engines feathered. None the less, the weather by this time was bad enough for the Dakota pilot to decide to turn back. As he did so, he caught another glimpse of the four-engined, triple-finned York through the heavily falling snow and thick cloud which appeared to have an indeterminate base of about 1,000ft. This was to be the last officially confirmed sighting of the lumbering York.

The first indication that something was amiss came via a signal from Pomigliano. MW126 should have landed there at 14.02hrs, and the signal timed at 15.50hrs requested news of the aircraft. Given who was on board, this query put the cat firmly among the pigeons. Searches were organised and carried out by the commands through whose airspace the York should have travelled. Twelve Dakotas searched between St Lô and Poitiers, while three more covered Poitiers to Marseilles, repeatedly for three days. More aircraft then joined the effort and in all 101 aircraft searched 61,500 miles for six days. All shipping and aircraft within the York's route were also warned to look out for it, but to no avail. Finally, on 26 November, all searches were abandoned.

Back on 14 November, in the tiny hamlet of Le Rivier d'Allemond — 1,265m above sea level in the Sept Laux mountains to the east of Grenoble — a snowstorm was raging. The inhabitants, being sensible French peasants, were firmly indoors when, soon after midday, the sound of aircraft engines was heard above the noise of the howling

Opposite: *Sir Trafford Leigh-Mallory, with a map of France in front of him, peering out of his personal aircraft* en route *to confer with Gen Montgomery at his HQ in Normandy. Leigh-Mallory probably took a similar interest in the progress of his flight down to Naples on 14 November 1944.*

Below: *Map showing the location of the village Le Rivier d'Allemond, high in the Alps, the site of Leigh-Mallory's fatal crash.*

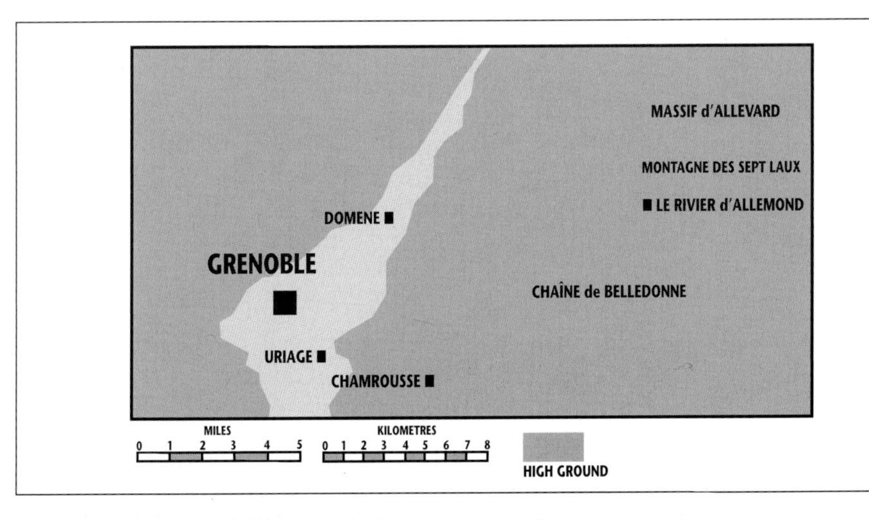

wind. Seraphin Mathieu, a worker at the local electricity station, went to his door. Just after the noise passed over the hamlet, M Mathieu heard an explosion and the sound ceased, leaving him to conclude that the aircraft had struck one of the nearby mountains. An extensive search was undertaken, both by the French gendarmerie and American military police, but both groups were thwarted by deep snow. It was not until June 1945 that the snows melted sufficiently to reveal the remains of the shattered York, its passengers and crew.

On 23 November 1944, a Court of Inquiry had assembled at HQ Fighter Command to look into the 'Accident involving York MW126 on 14 November 1944 — Passenger Air Chief Marshal Sir Trafford Leigh-Mallory'. All manner of scenarios were considered, and enemy black propaganda muddied the waters by claiming to have shot down the York with 'a new type of German long-range fighter'. But despite sitting until 29 November and taking evidence from 28 witnesses, the court found that there was insufficient evidence to determine the primary cause of the aircraft's disappearance. Nevertheless, the court found that dangerous weather conditions prevailed along the York's route, and that Sqn Ldr Lancaster's previous flying experience was insufficient 'to enable him to make a correct decision on meeting these weather conditions'. The court believed that the York should have either turned back or proceeded 'above cloud', not least because the pilot's view from a York 'is almost negligible both in rain and snow due to the design of the windscreen'.

There the matter rested until Saturday, 9 June 1945 when the Chief Inspector of Accidents at the Air Ministry received a signal from No 108 Repair and Salvage Unit at Marseilles to say that the wreckage of York MW126 had been found high in the French Alps near Le Rivier d'Allemond. On reaching the crash site, a party led by Gp Capt P. G. Tweedie, Deputy Chief Inspector of Accidents, found the wreckage strewn down the southeasterly face of a V-shaped crevasse at an altitude of 6,700ft. After examining the heavily twisted propeller blades, bent backwards over their reduction gears, Tweedie concluded that 'all four engines had been running under power at the moment of impact' with no signs of them being on fire.

The York had been flying in a north or northwesterly direction when it hit the top of the crevasse, tumbling back down the slope and disintegrating. In other words, the York was flying completely opposite to its scheduled track, which implies that its crew was well and truly disorientated. Assuming that the aircraft was flying between 7,000 and 8,000ft, this mountain was the first it would have encountered. But even if MW126 had cleared the peak, assuming it maintained track it would 'have flown into very much higher ranges' towering up to 13,000ft in the thin, mountain air. After four and a half hours at the site, the investigation party began the descent, leaving the remains of all aboard MW126 to be brought down by mule to be buried in the local village cemetery in 10 graves dug by 20 German prisoners-of-war.

A memorial service for Leigh-Mallory was held in Westminster Abbey on 28 June, but the day before the Chief Inspector of Accidents (CIA) forwarded a copy of Gp Capt Tweedie's crash report to the Director of Aircraft Accidents under a minute which referred, among other things, to the possible reopening of the Court of Inquiry. The CIA's

Opposite: *The circle shows the point 6,700ft up where the York hit the French Alps.*

opinion was that 'to do so would be a waste of time', a sentiment that was endorsed up to and including the Chief of Air Staff. With the ending of the war in Europe and the forthcoming general election, not to mention the dropping of atomic bombs on Japan, there were more weightier matters on people's minds.

So what caused MW126 to crash? In essence it flew into a mountain because its crew was woefully off track. The York should have been flown over the relatively flat land from Poitiers to Toulouse, down the Garonne valley to coast out at Sète, and then along the French Mediterranean littoral to Toulon before crossing the sea to Elba and Italy. But when MW126 crashed it was 250 miles off track, no small margin of error over a distance of approximately 360 miles. The aircraft probably got lost through the sheer inexperience of its flight crew. When OC No 511 Squadron, the unit that converted Sqn Ldr Lancaster on to the York at Lyneham, gave evidence to the Court of Inquiry, he recounted that he advised Lancaster to get more practice before the flight to southeast Asia and also to delay the trip if he possibly could to make at least one long cross-country flight. This was wise advice, given that Sqn Ldr Lancaster's recent navigational experience had been confined to flying-boats over the sea. But even more unfortunate was the fact that the second pilot and navigator were also ex-Sunderland men. With hindsight, picking two pilots and a navigator to fly an unfamiliar aircraft over 5,500 miles in winter was not the wisest piece of crew scheduling. Leigh-Mallory should have been given a crew picked from

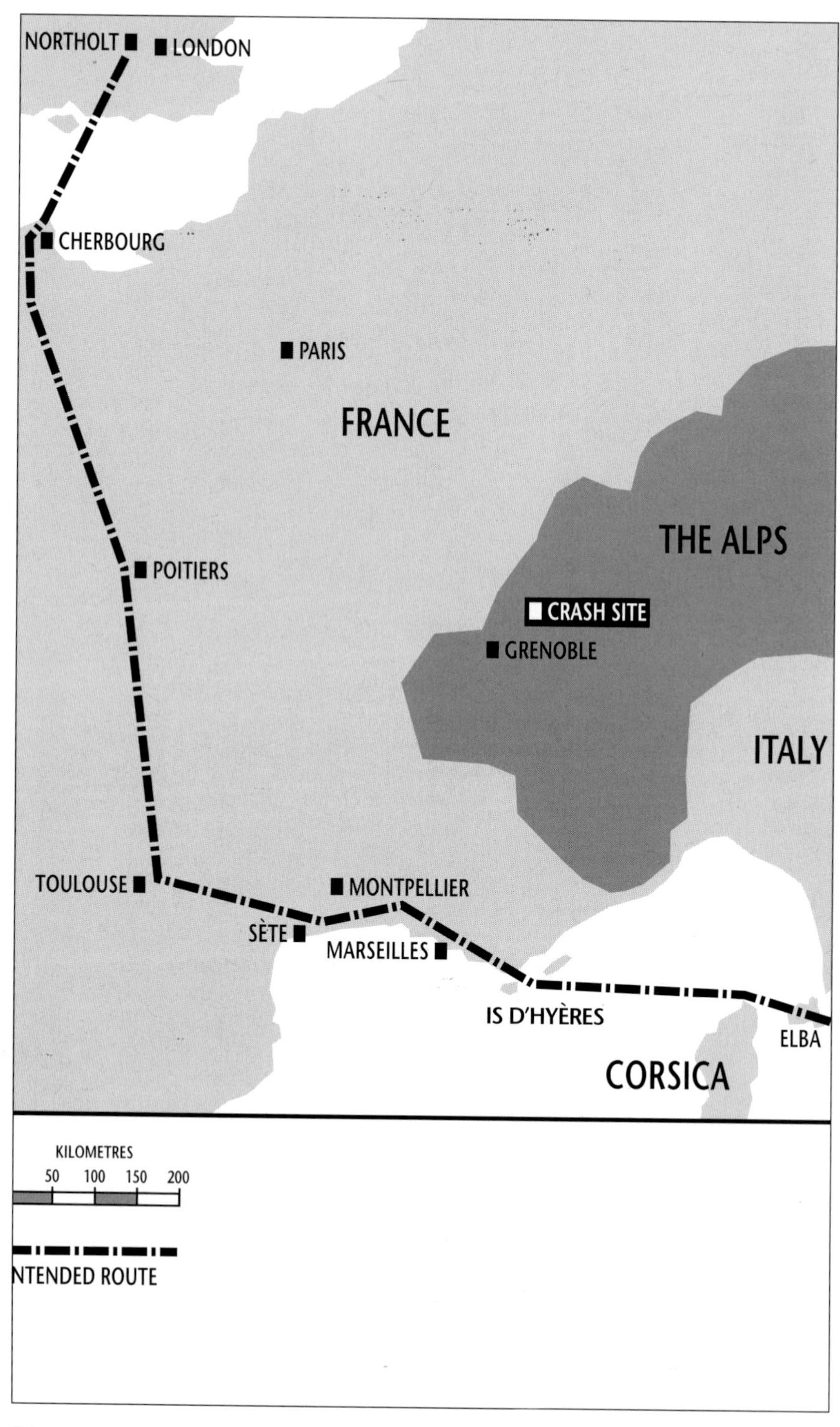
NORTHOLT
LONDON
CHERBOURG
PARIS
FRANCE
THE ALPS
POITIERS
CRASH SITE
GRENOBLE
ITALY
TOULOUSE
MONTPELLIER
SÈTE
MARSEILLES
IS D'HYÈRES
ELBA
CORSICA
KILOMETRES
50
100
150
200
NTENDED ROUTE

No 511 Squadron, the squadron most experienced in flying Yorks over long distances.

As it happened, OC No 511 Squadron had taken off from Lyneham to fly to Naples just a few hours earlier than MW126. After about an hour's flying his windscreen iced over and he flew virtually blind for the next 90 minutes, collecting ice on the spinners and radiator scoops. Whilst over France, he 'experienced heavy static' such that his wireless operator was unable to make contact with anyone and could only raise 'a continuous cackle through the static'. It was not until 15 minutes before reaching the southern French coast that the No 511 Squadron crew obtained a loop bearing.

It must have been pretty wretched on board MW126 that day as it lumbered towards its fatal meeting with the French Alps. The cockpit temperature in the Dakota that turned back over Normandy was reading -40°C, and in the opinion of the Dakota navigator who had been flying to and from North Africa and the Middle East for 18 months, the weather conditions on the morning of 14 November were the worst he had ever encountered. Maybe the flight crew aboard the unheated, uncomfortable York MW126 never knew how badly placed they were, or maybe they suspected but dare not admit it; it would have taken a pretty brave personal pilot who, on his first trip with the dominating Leigh-Mallory, would have said to the great man, 'I'm sorry, sir, we've cocked it up.' Or maybe Leigh-Mallory himself was the driving force. He was under some pressure to get out east, not least because his predecessor was being sent home expeditiously for having an affair with the wife of C-in-C India. Leigh-Mallory had cut down his embarkation leave so that he could get to Ceylon on 7 November, only to have his plans thwarted by last minute contraction of bronchitis. Maybe he insisted on pressing on because duty called, when all natural caution and airmanship training should have counselled turning back.

Since 1931, when a tri-motor Fokker crashed into the top of the Snowy Mountains *en route* from Sydney to Melbourne, hills have had a major impact on aviators. High ground is here to stay, and nearly 700 aircraft have flown into it since 1945. Some mountains appear to be more fated than others. An Air India Constellation flew into the 15,781ft high Mont Blanc in the Alps on 3 November 1950, killing eight crew and 40 passengers. In a chilling coincidence, an Air India Boeing 707 came to grief while carrying out the same approach as its predecessor into Geneva airport on 24 January 1966. Its captain should have started his descent from 19,000ft when over the summit of Mont Blanc, but he began to descend while the airliner still had five miles to run to the mountain. A horrified ground radar controller could only watch as the 707 blip suddenly stopped and then slid backwards, taking with it the lives of 117 people.

Yet, notwithstanding the claims made on behalf of all the wondrous gizmos designed to warn of upcoming terrain, the basics of flight safety are quite straightforward — all you have to do is aim at the ground and miss! Assuming that this is not so easy in a jet airliner travelling at eight miles a minute, on the face of it life should be much easier for the crew of a helicopter. After all, if you see a mountain wall looming up, you should be able to flare into the hover, turn around and beat a discretionary retreat. Would that it were so simple.

Left:
The map from the Court of Inquiry which traced the intended flightpath of Avro York MW126 on the first leg of its journey to Ceylon. MW126 was 250 miles off track when it crashed.

One of the best helicopters of all time is the Boeing-Vertol Chinook. Selected by the US Army Air Corps as its first standard helicopter transport in March 1959, the tandem-rotor turbine-engined Chinook had a fuselage interior large enough to carry the components of a Pershing missile. But it was the helicopter's ability to carry over 40 troops with full equipment that made this all-weather medium transport so popular with military forces around the world.

The RAF ordered the HC Mk 1 version of the Chinook in 1981. Subsequently, Boeing was authorised to update 32 RAF Chinooks to the HC Mk 2 standard which included updated hydraulics, stronger transmission, a new automatic flight control system and airframe reinforcements. The first HC Mk 2 to operate in Northern Ireland, ZD576, arrived at Aldergrove airport near Belfast on 31 May 1994. ZD576 belonged to No 7 Squadron at Odiham, Hants, and the helicopter and crew were the latest in a long series to serve on roulement in support of the security forces in Northern Ireland.

On Thursday, 2 June, the Chinook crew of four — two pilots and two crewmen — was tasked with a routine trooping flight; given the IRA threat, helicopter travel was the only safe way of getting to some parts of Northern Ireland. This task was completed without incident and the crew landed back at Aldergrove at 15.20hrs having flown for just under six hours. They were then briefed to ferry 25 people to Fort George, Inverness, but these were not just any old passengers. They included ten Royal Ulster Constabulary (RUC) Special Branch detectives, nine Army officers and five men and one woman attached to the Northern Ireland office, some of whom were said to be members of MI5. They were all on their way to Inverness for an annual gathering of experts from all branches of the security agencies involved in the intelligence war against paramilitary groups in Ulster.

The route to be flown would be as impressively scenic as any of the distinguished passengers could wish: flying north from Belfast across the Irish Sea to the Mull of Kintyre, up along the west coast of Kintyre past the RAF airfield at Machrihanish to Corran on the side of Loch Linnhe, over Fort William and down the Great Glen past Loch Ness to Fort George on the Moray Firth. During the pre-flight briefing, the crew was told that the forecast weather was generally suitable for the low level VFR route to Inverness, except in the Machrihanish area where it would be borderline with a 30% probability of 500m visibility and a broken cloudbase around 100ft. They left a copy of their route planning map behind, which together with evidence from various witnesses indicated that the crew's sortie planning was comprehensive and professional.

After a normal start-up, the passengers boarded and ZD576 left Aldergrove at 17.42hrs bound on a direct track for the Mull of Kintyre lighthouse, the first waypoint entered in the SuperTANS navigation equipment. The low level flight across the Antrim Hills and over the North Channel of the Irish Sea must have been very pleasant, and at 17.55hrs ZD576 made initial contact with Scottish Military air traffic. The helicopter was then seen about 3nm southwest of the Mull of Kintyre lighthouse by a yachtsman, who estimated that it was flying straight and level at about 200-400ft below cloud.

Around 18.00hrs, several witnesses near the lighthouse heard the mighty Chinook, 99ft long with its great rotors churning, approaching at 150kt from the southwest.

Opposite: *An RAF Chinook HC Mk 2 working for the UN in Bosnia.*

The weather in the area was poor, with cloud and hill fog extending from below the level of the lighthouse (300ft above sea level) to beyond the 1,400ft summit of the highest point on the southern tip of the Mull, the great mass of Beinn ne Lice. The witnesses did not have sight of the Chinook because visibility around the lighthouse was 400-500m, but shortly afterwards they heard a dull thump followed by a loud whooshing and whistling. An area of flame and smoke then glowed on the foggy hillside.

ZD576 crashed into rising ground 500m east of the lighthouse. Despite breaking up on initial impact, the great helicopter briefly became airborne again while continuing to fall apart: the main wreckage came to rest some 300m from initial impact point. There was an extensive post-impact fire but no one on board would have known about it. High speed impact forces killed all 29 people on ZD576 instantly, and no practicable crashworthy design features could have saved them.

The subsequent RAF Board of Inquiry investigation was hampered by the lack of a cockpit voice recorder or accident data recorder, neither of which had been fitted during the Chinook HC Mk 2 mid-life update. Nor were there any eyewitnesses to the crash itself. Given the background of those on board, a close eye was kept out for the possibility of sabotage, but there was no evidence of any damage caused by small-arms fire, surface-to-air missile or on-board explosive. The helicopter did not suffer any major structural or technical failure prior to impact, and interference with controls or navigation equipment was not a factor in the accident.

Turning to the forecast weather, the Board of Inquiry felt that while there was an isolated risk of conditions deteriorating to preclude VFR in the Machrihanish area, the Chinook captain was justified in attempting a VFR transit provided he was prepared to fall back on an IMC contingency plan if necessary. The weather over the Mull of Kintyre area was very poor at the time of the accident and certainly well below the minima required for low level VFR transit.

When tragedy struck, the crew had been on duty for 9hr 15min, including some six hours' flying time. If they were to return to base that evening, they would have needed the personal authorisation of OC Aldergrove. Although there was no evidence of fatigue playing any part in the accident, the crew may have felt under pressure to complete the sortie to Inverness as quickly as possible.

Although the barometric altimeter was set correctly, the radar altimeter setting procedures adopted by the crew probably denied them any effective ground proximity warning in the circumstances. But that was a bit of a red herring. Such an experienced and respected crew, equipped with accurate SuperTANS and GPS navigation equipment, should have known where they were in relation to the high ground up ahead, and they could see the deteriorating weather for themselves. Yet, the 30-year-old captain in the left-hand seat, and the 28-year-old co-pilot doing the flying from the right, did not appear to have reduced speed or turned away from the high ground on encountering poor visibility and lowering cloudbase. As it happened, the crew changed the waypoint on their navigation equipment some 15-18sec before impact, which was consistent with them continuing VFR up the Kintyre coast to Corran having sighted the Mull. All the Board of Inquiry could determine was that if the penny did drop, the pilots selected a rate of climb inappropriate to safe overflight of the Mull of Kintyre.

The RAF Inquiry was therefore left withthe impression that the Chinook crew, when faced with expected poor weather on approaching the high ground of the Mull of Kintyre, neither maintained safe VFR flight nor made a safe transition to IFR flight. Their actions were therefore in direct contravention of the rules for flight in both VMC and IMC. Reluctantly, RAF higher authority concluded that the lack of situational awareness of the two pilots — the crewmen were absolved of any blame — was the direct cause of the accident. In ignoring one of the most basic tenets of airmanship, which is never to try to fly visually below safety altitude unless the weather conditions are unambiguously suitable for operating under VFR, both pilots were found to have been negligent.

Despite a susequent inquiry by a Scottish sheriff which said that it was unclear why the Chinook crashed, the RAF stuck by its original finding that a perfectly serviceable aircraft had been flown into a mountainside. The loss of Chinook ZD576 marked the worst loss of life in a single RAF accident since 1972, and it stimulated retrofitment of accident data recorders and cockpit voice recorders into the Chinook fleet by 1998. It was doubly ironic that the crash occurred *en route* to a relatively relaxing conference for a body of dedicated security and intelligence personnel who, by the very nature of their profession, had survived everything that some pretty ruthless opposition could throw at them. Saddest of all, men like Assistant Chief Constable Brian Fitzsimons, head of the RUC Special Branch and the most senior officer to die in the crash, did not live to see the easing of tension in Northern Ireland for which he and his colleagues on Chinook ZD576 had worked so hard.

Whatever the cause of the loss of Chinook ZD576, it was but one further example of the old truism that in any battle between aircraft and high ground, the aircraft invariably comes off second best.

6 These Foolish Things

Just after noon on 31 July 1993, BAe146-300 (G-UKSC) was parked on stand No 10 at Gatwick with the crew at its designated positions and passengers boarding for a flight to Edinburgh. The pilot sitting in the right-hand seat looked out of his window and saw a driverless baggage tractor, travelling at walking pace, approaching the nose of the 146 at right angles to the aircraft's centreline. As the tractor approached, a member of the ground staff was seen to make a spirited but unsuccessful attempt to prevent the vehicle from hitting the aircraft just forward of the nose undercarriage. After the impact, which caused no injuries, further boarding was stopped and all passengers returned to the departure lounge.

Various airport luminaries attended the scene, where initial visual inspection revealed that the tractor engine was still running and its automatic gear selector was in the 'drive' position. The aircraft commander assessed that the tractor driver had deserted his vehicle with the engine running, in gear and pointing towards the 146. An investigation carried out by the ground handling company confirmed the commander's assessment. Even though the driver had left the vehicle unattended with the handbrake on, tests confirmed that

Below: *The crunched nose of G-UKSC after the driverless baggage tractor had driven into it.*

after a period of between 4-5min, sufficient pressure accumulated within the automatic transmission, with the vehicle in gear, to overcome the handbrake. In the wake of this accident, the ground handling company fitted a 'deadman's pedal' to all similar vehicles which cut off the fuel supply to the engine when foot pressure was released.

The vision of driverless vehicles wandering around airports verges on the hilarious, but accidents such as that which befell the 146 can be expensive in terms of repair effort and productive time lost. Inadvertent ground collisions can also have an impact that make them no laughing matter.

At 18.39hrs on 27 December 1979, a Pan American Boeing 747-121 freighter (N771PA) touched down on Runway 23 at London Heathrow. The only occupants were the commander, co-pilot and flight engineer who were engaged on a scheduled international cargo flight from John F. Kennedy International airport, New York (JFK). Following a heavy touchdown, the commander applied reverse thrust. Shortly afterwards the flight engineer reported that the exhaust gas temperature was approaching limits on No 4 engine, whereupon the commander began to reduce reverse thrust on all engines. As he did so, he felt No 4 thrust lever give a sudden jerk and thereafter become immovable. After reverse thrust had been cancelled on the other three engines, it became apparent that No 4 was not delivering reverse thrust despite the position of its thrust lever. All the instrument indications were that No 4 had run down and flamed out. Whilst N771PA was moving slowly clear of the runway to the right, the co-pilot observed a large fire in the region of No 4 engine although there was no fire warning indication on the flightdeck. ATC had by then activated the crash alarm. The crew carried out the engine fire drill and stopped the 747 clear of the runway. The other engines were shut down, the fire was rapidly brought under control by the airport fire service, and the crew got out safely.

Subsequent investigation showed that, at touchdown, the No 4 pylon forward bulkhead, which supported the front of the engine, began to break free of the pylon because of weakening by fatigue and other pre-existing damage. The resulting downward movement of the engine during the landing roll ruptured the engine fuel feed pipe and several other connections with the engine, including engine monitoring and fire warning circuitry. The high volume of fuel issuing from the ruptured fuel pipe caused a severe fire to develop under No 4 pylon. The fire continued until the crew physically observed the fire as their aircraft was turning off the runway, at which point the fire drill was carried out. This closed the fuel shut-off valve in the wing.

The investigation noted that No 4 engine was involved in a collision with a baggage container at Chicago airport three years previously. This collision may have been the initiating factor in the chain of structural breakdown which culminated in the separation of the pylon forward bulkhead — the direct cause of the accident — all that time later on the other side of the Atlantic.

Silly little errors and omissions can have just as catastrophic an effect as big ones. At 01.04hrs on 21 January 1985, a Lockheed 188 Electra (N5532) belonging to Galaxy Airlines took off from Reno-Cannon International airport, Nevada, bound for Minneapolis. Less than 30sec later the first officer requested permission

Opposite: *A Lockheed Electra, similar to that which crashed at Reno on 21 January 1985.*

to land, reporting 'heavy vibration'. Cleared by the controller for a return to the airport, the Electra began a left turn, climbing to an estimated 200-250ft above the ground. Its undercarriage was retracted when N5532 crashed approximately one mile from the end of the runway and about half a mile to the right of the extended centreline. All but a single passenger among the 71 persons aboard were killed, including the crew of five. Several vehicles in a dealership lot were also destroyed in the accident, which occurred in darkness but clear weather.

This was a classic case of the crew 'rushing' to a fatal extent. They were running to a tight deadline, with a departure for Seattle slated less than 90min after the Electra's arrival at Minneapolis. Analysis of the CVR tape revealed improper use of the 'before-start checklist', with certain items omitted, possibly to save time. The Electra even started to taxi away from the gate with the air-start hose still attached. This hose passed high pressure compressor air, via an inlet on the starboard leading edge, to turn the turbine blades of No 3 engine: once No 3 got going, it could cross-feed air to start the other engines. When N5532 began to move, the hose was pulled taut, preventing the ground handler from disconnecting it. Her supervisor removed it for her, but neither could remember closing the air-start hose access door.

From background noise on the CVR, impact damage on the open door latch, and statements from other pilots who had previous experiences with open air-start access doors, the NTSB concluded that this was the source of the vibration mentioned by the first officer. Although it would not have affected N5532 aerodynamically, this vibration led to a breakdown in crew co-operation. The captain tried to establish the cause of the noise and fly the aircraft, and he was unsuccessful on both counts. Apparently believing that the trouble lay with the engines, he ordered that all four power levers be retarded. It would have been wiser to leave them as they were, climb to height, and check the engines individually once at a safe altitude. As it was, the significant reduction in power bled off air speed and led to a stall. The co-pilot, in his haste to respond to the demands of his considerably older and more experienced commander, plus ATC requests, neglected to monitor height or speed adequately. His eventual call of 'a hundred knots' came too

late to prevent the crash, despite the application of full power. And all this came about because of something as silly as an access door left open.

But if there is one thing that pilots should never mess about with, it is their fuel reserves. Capt Balsey DeWitt found this out the hard way on 2 May 1970 when in command of a nearly-new DC-9-33F (N953F) operated by Overseas National Airways. Designated Flight 980, the DC-9 was scheduled to leave JFK for St Maarten's Juliana airport on the Netherlands Antilles, a small group of islands in the West Indies. *En route* weather for the 1,400nm trip was expected to be overcast and showery, with scattered thunderstorms. Selecting St Thomas airport in the US Virgin Islands as the alternate, the crew filed an IFR flight plan direct to St Maarten at FL290, with a flight time of 206min. The DC-9 carried 28,000lb of fuel, enough to fly to St Maarten, divert to St Thomas and hold there for 30min if necessary, and still have 900lb to spare.

When the 57 passengers had boarded, the flight crew discovered that the public address system did not work. Although this would be a nuisance during the three and a half hour flight, it was not a show stopper. The fault appeared to lie with the flightdeck microphone, and the PA system within the passenger cabin was serviceable.

The DC-9 got airborne at 11.14hrs, and cruising flight was established at FL290. Some 750nm southeast of New York, Flight 980 encountered a thunderstorm but it got round the attendant turbulence by doglegging and descending to FL270. Thunderstorm activity persisted and with 400nm to run to St Maarten, the crew requested a further descent to FL250. In contact with San Juan Air Route Traffic Control Centre, Puerto Rico, when the DC-9 was 180nm northwest of St Maarten, the crew noted that they had 8,600lb of fuel remaining. They estimated that they would arrive at St Maarten at 15.00hrs with 6,000lb on board — that was already 400lb lower than the reserve calculated before take-off.

Some 15min later, San Juan advised Capt DeWitt that the weather at St Maarten had deteriorated and was now below the landing minima of 600ft cloudbase and 3km visibility. The captain's response was faultless: he asked for a clearance to St Thomas airport at FL210 and at 14.46hrs, less than 15min from Juliana, he diverted the DC-9 to the Virgin Islands.

Five minutes later it started to get silly. Juliana's tower

Above:
An Overseas National Airways DC-9.

controller reported that the weather had improved, so Capt DeWitt decided to try a landing at Juliana after all. At 14.51hrs, San Juan Control passed Flight 980 a new clearance to St Maarten. The crew had 5,800lb of fuel at this stage and estimated that they would arrive at the Juliana terminal with 4,400lb in the tanks. At 15.08hrs Flight 980 reported over the Juliana NDB at 2,500ft; the controller advised that there was now scattered cloud at 800ft, broken cloud at 1,000ft and visibility fluctuated between 3-5km.

While First Officer Harry Evans worked the radios, Capt DeWitt flew an NDB approach with the undercarriage lowered, though every minute flown with the draggy gear down increased fuel consumption. The crew finally became visual at 15.15hrs. The airfield lay immediately ahead of them but it was too late to align the DC-9 properly with the single east/west runway that was less than 5,000ft long. Capt DeWitt applied power and made a level left-hand turn to try and fly a visual circuit below the cloud.

A second approach four minutes later had to be abandoned because of a heavy shower on base leg. The third time round, the commander succeeded in getting the alignment right but when the runway came into view, Capt DeWitt realised that he was too high and too close for anything other than a glide approach. Forced to go round yet again, the commander briefly discussed the situation with his navigator, Hugh Hart, and then decided to divert to St Thomas after all. It was 15.31hrs when the controller cleared the DC-9 direct to St Thomas — 110nm away — at 4,000ft.

Re-establishing contact with San Juan Control, Flight 980 was informed that it would be assigned a higher altitude shortly; there was a slower aircraft only 10nm ahead. In the brief climb to 4,000ft, all three members of the flight crew were suddenly alarmed to see the fuel gauges give erratic readings; as the DC-9 levelled out at 4,000ft, the totaliser stabilised momentarily at only 850lb. Shortly afterwards, San Juan asked, 'What altitude are you requesting to St Thomas?' A deeply worried commander responded, 'Anything you've got that's higher — I'm a little short on fuel and I've got to get up.'

Flight 980 was cleared to FL120, whereupon Capt DeWitt gently coaxed the DC-9 into a climb, using less power and lower airspeed than normal in attempt to conserve precious fuel. The gauges were still behaving erratically and Navigator Hart suggested that it might pay to divert to the airport on the US island of St Croix, which was 10nm closer than St Thomas. It seemed a good idea so the captain turned left for St Croix. As the airliner broke out above cloud at about 7,000ft, Capt DeWitt became even more uneasy. If they were so short of fuel that they had to ditch, he did not want to flameout in cloud and have to make a deadstick approach IMC. He wanted the time to set up a ditching, which meant keeping the sea in sight. The crew agreed, so clearance was sought from San Juan to descend again. This request was approved but as the DC-9 passed through 5,000ft, the captain decided that it might be best to get down close to the water and find a place to ditch there and then. Calling San Juan himself, Capt DeWitt told the controller, 'OK, I may have to ditch this aircraft — I'm now descending to the water.' A rather surprised purser was called forward and told to brief the passengers to don their lifejackets and to prepare generally.

San Juan was given the likely ditching position and the crew was assured that rescue efforts were being put in hand. At 15.47hrs N953F was down at 500ft just below the base of the cloud. The grey-swelling sea was plainly visible but the rain was heavy, and horizontal visibility was down to 600m. Fuel gauge readings showed that the Pratt & Whitney JT8D engines were just about running off the fumes in the tanks, and First Officer Evans told San Juan, 'We're ditching.'

Levelling off, Capt DeWitt positioned his aircraft directly over the crest of the swell and flew a pretty good steep descent to improve his depth perception. Down at 20ft with speed decaying, he lowered full flap as soon as the fuel pressure warning lights started to flicker. Soon afterwards, both engines flamed out. With no working microphone to communicate rearwards, the pilots could only flash the 'Fasten Seat Belt' and 'No Smoking' signs to warn that ditching was imminent.

Back in the passenger cabin, Navigator Hart was helping the purser remove one of the aircraft's 25-man inflatable life-rafts when he heard the engines spool down. Seeing the 'Fasten Seat Belt' signs flashing, he suddenly realised that they were about to ditch. In the cabin, several passengers were still standing while putting on their lifejackets, and at least five others did not have their seat belts fastened. Desperately shouting for everyone to sit down, the navigator and purser quickly occupied the aft-facing jump seat on the forward cabin bulkhead but neither had time to fasten his seat belt.

Quite a few people at the back did not appreciate the urgency of the situation until the airliner struck the water. Some were unprepared for the severity of the impact — they never braced themselves for the 'crash position' — and passengers who were still standing or did not have their belts fastened were flung violently forward. At least six more were injured when their seat belt fastenings failed. Yet the DC-9 itself remained essentially intact and only the stewardess from among the crew members was injured. Immediately the aircraft lurched to a stop, the navigator and purser tried to open the port side forward main door, only to find it jammed. A steward opened the galley loading door on the starboard side, through which passengers started to scramble. The navigator and purser went to help this steward retrieve the life raft from beneath galley debris. They had just been joined by First Officer Evans from the flightdeck when the life-raft inadvertently inflated, pinning Evans against the bulkhead and blocking the others' way back into the main cabin. They did the only thing possible and jumped into the sea via the galley door, as did Evans as soon as he was able to extricate himself.

The DC-9 by this time was rapidly taking in water. Fortunately, the passenger sitting next to the aft overwing exit on the starboard side had opened it. Being a seasoned air traveller, he made it his business to sit next to an emergency exit whenever possible, and to make a mental note of its operation. He and most of the able-bodied passengers, all wearing lifejackets, quickly left the aircraft through this exit.

Alone on the flightdeck, Capt DeWitt saw the inflated life raft blocking his way to the main cabin so he scrambled through the windscreen hatch. Swimming back to the port side overwing exits, he opened them from the outside. After assisting two passengers out, he looked inside for more but he could see none. By this time, the DC-9 was close to sinking and it was imperative that those in the water got clear. N953F, an aircraft far more sinned against than sinning, finally sank beneath the waves less than 10min after ditching.

Although all five life-rafts on board went down with the airliner, the inflatable escape chute for the galley door had dropped into the water while the crew was escaping. Survivors in the water clung to this and before long, a US Coast Guard HU-16 Albatross amphibian and then a Shorts Skyvan arrived overhead from Puerto Rico to drop life-rafts. An hour and a half after the ditching, two Coast Guard HH-52 helicopters and then a US Navy Sea King came to uplift 37 people between them. Finally, an hour later, a US Marine Corps Sea Knight winched up the three remaining survivors, the last to leave the water being First Officer Evans. When a tally was finally made, it was found that 22 passengers (including two children) and Stewardess Margaret Abraham were missing.

The DC-9 sank in 5,000ft of water but it did not need a major salvage operation to tell NTSB investigators that the accident resulted from inadequate fuel management. Yet the original planning had not cut things fine. The flight plan to St Maarten had complied fully with Federal Aviation Regulations which demanded a 10% reserve, fuel to get to the St Thomas alternate and 30min holding fuel. In fact, N953F carried 900lb fuel over and above this.

Clearly, if the airliner had continued to San Juan when Capt DeWitt first decided to divert at 14.39hrs, Flight 980 would have landed with fuel to spare and this accident would never have happened. In the event, Capt DeWitt was persuaded against what should have been his better judgement to turn back by Juliana tower's advice that conditions at St Maarten had lifted above minima and showed an improving trend. The commander of an airliner has an obligation to deliver paying passengers to where they want to go, but he has an even greater obligation to get them down on the ground in one piece. Capt DeWitt had already jinked to avoid thunderstorm activity *en route*, so N953F was never going to meet its planned flight time of 206min. At 14.51hrs, when the captain reversed his decision to go to San Juan, the DC-9 had already been in the air for 217min. Irrespective of any optimistically seductive words from meteorologists or air traffickers on the ground, the rules in this instance were quite clear. Capt DeWitt should only have changed his plan to divert to St Thomas provided he had enough fuel to fly to St Maarten, carry out an NDB approach and then, if it proved necessary, divert to the planned alternate. In opting to turn back to St Maarten without such fuel reserves, the commander of Flight 980 made his first major error of judgement.

At this stage, according to Capt DeWitt, he had 5,800lb of fuel remaining in the tanks. As his revised ETA for St Maarten was 15.05hrs, and if a straight-in approach was possible, he believed he would be on the ground with just under 5,000lb left, which kept its prescribed fuel reserves intact. But Flight 980 did not reach St Maarten until 15.15hrs, and most of these extra 10min were flown at a fuel-hungry 2,500ft. By the time it reached the St Maarten NDB, the DC-9 was down to a critical endurance of 33min, necessitating an immediate diversion to the alternate if an approach and landing could not be carried out promptly. After the first approach was abandoned, only 29min remained to fuel exhaustion. But instead of diverting, the commander began two tight, low altitude circuits in an effort to land. These took another 11min, the DC-9 being in the fuel-gobbling landing configuration throughout. Some 1,400lb was consumed in the process, leaving only 2,200lb in the tanks when the decision was finally made to divert to St Thomas.

This was slightly more than the fuel allowed for the diversion in the

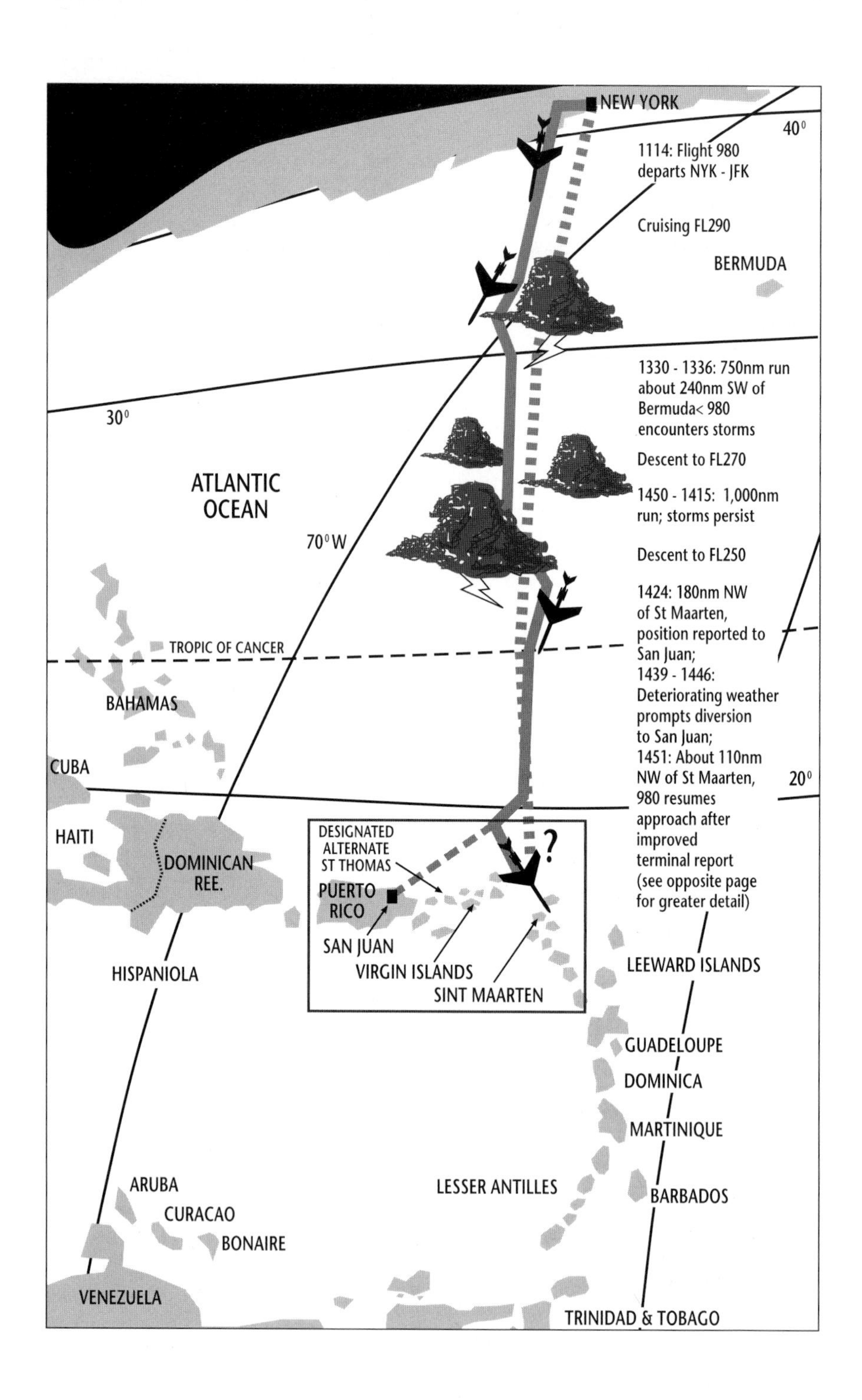

NEW YORK
40°
1114: Flight 980
departs NYK - JFK
Cruising FL290
BERMUDA
1330 - 1336: 750nm run
about 240nm SW of
Bermuda< 980
encounters storms
30°
Descent to FL270
ATLANTIC
OCEAN
1450 - 1415: 1,000nm
run; storms persist
70°W
Descent to FL250
1424: 180nm NW
of St Maarten,
position reported to
San Juan;
TROPIC OF CANCER
1439 - 1446:
Deteriorating weather
prompts diversion
to San Juan;
BAHAMAS
1451: About 110nm
NW of St Maarten,
980 resumes
approach after
improved
terminal report
(see opposite page
for greater detail)
CUBA
20°
HAITI
DOMINICAN
REE.
DESIGNATED
ALTERNATE
ST THOMAS
PUERTO
RICO
?
SAN JUAN
VIRGIN ISLANDS
SINT MAARTEN
HISPANIOLA
LEEWARD ISLANDS
GUADELOUPE
DOMINICA
MARTINIQUE
ARUBA
LESSER ANTILLES
BARBADOS
CURACAO
BONAIRE
VENEZUELA
TRINIDAD & TOBAGO

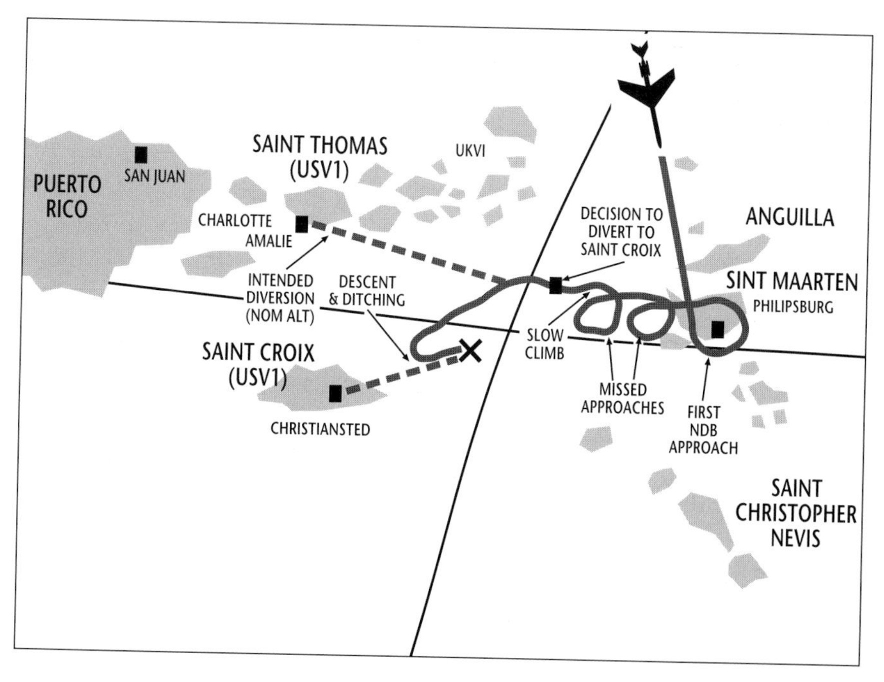

flight plan, but it was a figure based on zero wind conditions and an optimum flight profile. But despite the DC-9's critical fuel state, when the captain told Juliana tower that he was diverting he did not request an economical altitude. Nor did he tell San Juan of his parlous fuel state, so the controller let the DC-9 bimble along until another airliner was out of the way before approving the climb to FL120. And even at this stage, the commander did not use full power. He chose a low power, low airspeed climb to conserve fuel, but this was much less efficient in terms of distance covered than a full power climb. Any opportunity the airliner had to reach St Thomas or St Croix disappeared at this time.

Notwithstanding the problems in communicating between flightdeck and main cabin, the ditching itself was well planned and most successfully accomplished under adverse conditions. Capt DeWitt demonstrated exceptional skill in handling the aircraft, and the leadership shown by all flight crew members after the DC-9 hit the water did much to minimise the subsequent loss of life while awaiting rescue. Coincidentally, in similar waters near San Juan in 1952, the ditching of a Pan Am DC-4 was also marked by the flight crew's failure to warn the cabin attendants of the situation, which prevented them from preparing passengers for a watery landing. Today's routine procedure of briefing passengers on the location and use of flotation equipment and emergency exits stemmed from this DC-4 accident. As water covers 70% of the earth, it is prudent to know your ditching drills. The loss of Flight 980 also showed that there may be no flight attendant to hand when you need one the most; it can therefore pay to know how to open the nearest emergency exit.

Opposite and above: *The path taken by Flight 980 on 2 May 1970, including storm avoidance and diversion routeing.*

Having demonstrated flying ability of a high order at the very end, it was a great pity that earlier, Capt DeWitt had been so preoccupied with getting his passengers down at St Maarten that he 'lost the plot' concerning the aircraft's remaining fuel reserves. The extended routeing to avoid thunderstorms, the dicey weather at St Maarten with its single, dinky runway and no precision approach aids, all argued against cutting fuel reserves too fine. To have saved many lives in the end was admirable, but to get into the situation where any had to be lost at all was lamentable. To watch two 14,000lb thrust engines wind down because you have run out of fuel for no good reason is the ultimate aeronautical foolishness. When engines fail in the air, unlike on the motorway, there is no pulling into the side to await assistance. It should never be forgotten that in the air, lift is a gift but thrust is a must.

7 The *Vincennes* Incident

At 10.15hrs Iranian time on 3 July 1988, Iran Air Flight 655 began its take-off roll down Runway 21 at Bandar Abbas airport in southern Iran. In command on the flightdeck of the Airbus A300B2-202 (EP-IBU) was Capt Mohsen Rezaian, a US-trained pilot aged 38 who began his civil aviation career on Boeing 737s in 1975. He had met up with 31-year-old First Officer Kamran Teimouri, a commercial pilot for four years, and 33-year-old Flight Engineer Mohammad Reza Joze at Tehran airport at 05.30hrs that morning for a routine briefing prior to the first leg of their day's work to Bandar Abbas. From there they would take the A300 to Dubai, returning to Tehran that evening. It should all have been a piece of cake: the weather forecast was good and during the previous two months, Capt Rezaian and his co-pilot had criss-crossed the Persian Gulf on the Tehran-Dubai route 24 times between them.

Unfortunately, if the routine was normal the background circumstances were not. Although the Iranian Airbus was the 67th civilian aircraft to pass down airway Amber 59 — which crossed the neck of the Gulf near the Straits of Hormuz — over the preceding calendar month, Flight 655 was to enter an arena wherein life was getting hectic. For

Below: *An Iran Air A300B2-202, identical to the Airbus shot down over the Persian Gulf on 3 July 1988.*

nearly eight years Iran and Iraq had been at war, but ever since its abortive 1987 drive to take the important Iraqi city of Basra, Iran had suffered a nearly unbroken string of political and military defeats. Given the widespread Iranian assumption that the United States, or 'Great Satan', was backing Iraq, all US forces in and around the Middle East were placed on their highest state of readiness in case of a possible Iranian attack to coincide with American Independence Day on 4 July.

Capt Rezaian would have been as aware as anybody that his aircraft was not going to operate in 'normal' airspace. Integral to his crew's briefing at Bandar Abbas would have been notification, if appropriate, of any red alert status for that area of the Gulf. These alerts were sent out by the Iranian military, giving warning of any impending attacks on shipping or sparring with Western navies, when all air traffic clearances were withheld. As it was not unknown for airliners *en route* to be recalled, there was a standing order that the civilian distress frequency should be monitored continuously. But on 3 July there were no red alerts. Flight 655, with its crew of 16 and 274 passengers, lifted off from Bandar Abbas for the 125-mile trip to Dubai with no idea of what was taking place below it off Qeshm island.

At about the time the Airbus left Tehran, a 'Knox' class frigate, USS *Elmer Montgomery*, was observing 'small boat activity' in the Straits of Hormuz. Within an hour it was reporting back to the commander of Destroyer Squadron 25, aboard his flagship off Bahrain, that there was heavy Iranian radio traffic and explosions to the north. The commander's response was to order his force anti-aircraft command ship, the cruiser USS *Vincennes*, to move in to support the *Montgomery*. By 10.15hrs Emirates time — 09.45hrs Iranian time — the 'Perry' class frigate USS *John H. Sides* had joined the pack, and *Vincennes'* Sikorsky SH-60B Seahawk LAMPS 3 (Light Airborne Multi-Purpose System Mk III) helicopter was scouting some 10 miles north of the *Montgomery* to investigate the Iranian small–boat activity.

Flight 655 was still at Bandar Abbas terminal when the *Vincennes'* helicopter reported that it was being fired upon by the small boats. This was not wholly unexpected; the boats were manned by the fanatical Revolutionary Guard, and a Danish tanker had come under fire from similar gunboats only the day before. At 10.20hrs the commander of Destroyer Squadron 25 ordered the captain of the *Vincennes*, William Rogers III, to take tactical control of the *Montgomery*, and the pair moved into position to drive the gunboats off. Working up speed from the south, *Vincennes* sighted the small boats within three minutes. They were faster than the *Vincennes*, and they did not leave the area when the cruiser began firing. After requesting permission, *Vincennes* engaged the boats which were closing at high speed. Bullet damage was subsequently found on the cruiser's starboard bow.

Vincennes had two Mk 45 5in guns and two Phalanx close-in weapon system defensive guns; the latter provided last ditch-defence against incoming sea-skimming missiles and therefore could not engage surface targets. On 3 July, the forward 5in gun mount failed after one minute, leaving only one gun mount to engage up to four surface contacts. *Vincennes* had to be manoeuvred violently (30° turns at 30kt) to protect its blind side from attack while keeping the 5in mount bearing on the gunboats. This engagement lasted 17min, during which time the ship fired 72 rounds.

Three minutes after Capt Will Rogers had received permission to engage the small Iranian boats with gunfire, Bandar Abbas tower cleared

Capt Rezaian 'to destination Dubai via flight plan route, climb and maintain Flight Level 140, after take-off follow Mobet 1 Bravo departure squawking 6760'. Mobet was the midway reporting point on Amber 59, and the squawk was Flight 655's electronic signature code which would identify it as a civilian flight on air traffic controllers' radar. As he taxied out, Capt Rezaian may well have glanced at the Iranian Air Force F-14 Tomcat fighters that only three days earlier had moved from Bushehr air base to the military section of Bandar Abbas adjacent to the civil airfield.

Vincennes' crew was busy tracking an unidentified aircraft 62 miles to the west, later identified as an Iranian Air Force P-3 Orion maritime patrol aircraft. Then, barely 30sec after it took off at 10.47:37hrs, Flight 655 was picked up by *Vincennes'* SPY-1A radar. Enter technology. The *Vincennes* was one of eleven Aegis-equipped 'Ticonderoga' class cruisers each built at a cost of $1.2 billion. As the name implies — Aegis was the mythical shield of Zeus — each cruiser was originally built to defend carrier battle groups from air threats. Capt Rogers ran the show from below decks in the Combat Information Centre (CIC) dominated by four huge blue computer screens. He and his battle staff could see nothing other than what the software showed them, but that should have been sufficient because the theoretical capability was awesome. The Aegis system comprised the SPY phased-array radar which could track more than 100 aircraft, surface ships, submarines, missiles and torpedoes simultaneously. All showed up as white symbols on one of the four blue screens, each symbol being of a particular shape identifying the object as an aircraft, missile or whatever. Computers then directed the firing of up to 88 General Dynamics STANDARD SM-2MR surface-to-air missiles in one engagement plus other weapons at every form of threat. The SPY radar could supposedly spot a basketball at 150 miles and a high altitude aircraft at over 1,000 miles, but head-on, it could not tell an Airbus from an F-14 interceptor.

For a given target, the large-screen display on the *Vincennes* showed only a vector, giving course and speed (not altitude as the display was two-dimensional). Thereafter, it was up to CIC personnel to decide what the target was, and typically they began by tagging the Airbus as an unknown, with a track number or 'TN'. From start to finish, *Vincennes'* CIC identified Iran Air 655 as unknown TN4131 but, understandably in a hostile area such as the Persian Gulf, unknown often equated with 'probably hostile'.

Vincennes' automatic detection and tracking operator identified Iran Air 655 when it was 47 miles to the north. Thereafter, TN4131 closed steadily on a constant bearing, which was not surprising given that it was flying (albeit off-centre) down Amber 59 and the cruiser was positioned under Amber 59. In the CIC, the Airbus could easily have been mistaken for an Iranian F-14 flying directly towards the *Vincennes* on a classic attack path.

Discrimination between civil and military traffic should have been provided by IFF interrogation. In the Gulf theatre, Mode 2 was associated with Iranian military aircraft and Mode 3 with civilian. As Iran Air 655 was squawking a clear civilian Mode 3C transponder return throughout its flight, it should have been safe, but as soon as it climbed off the Bandar Abbas runway, the *Vincennes'* identification supervisor tagged TN4131 as 'Mode 2-1100, breaks as an F-14'. Nothing in the ship's data banks pointed to such an identification; this was simply a mistake which then took on a life of its own.

It should never be forgotten that much else was happening around this time. Capt Rogers was preoccupied with defending his ship, as well as his

LAMPS 3 helicopter, against Revolutionary Guard gunboats. When one of his 5in guns jammed and the *Vincennes* began its series of 30°/30kt turns, kit and documents were scattered over the control room and life became even less conducive to measured deliberation.

Other challenges then had to be dealt with. At 10.48hrs, *Vincennes* began challenging the P-3 Orion patrol aircraft:

Vincennes: 'Iranian Papa 3 on course 085 speed 27, this is United States warship bearing 086, range 64 miles. Request you state your intentions, over. Stay clear of my unit, over, stay clear of my unit.'

Orion: 'US warship, this is Iranian Papa 3, our intention is search mission. We keep clear of your unit.'

Vincennes: 'Iranian Papa 3, this is US warship, stay 50 miles from my unit, repeat 50 miles.'

A minute later, while *Vincennes'* forward 5in gun fired away, her anti-air warfare co-ordinator ordered challenges to the suspected F-14 to begin, on both 243MHz (the military distress frequency) and 121.5MHz (its civilian equivalent). The first challenge went out on the military frequency, which the airliner would not have had dialled up.

Vincennes: 'Unidentified Iranian aircraft, you are approaching United States naval warship in international waters, request you state your intentions.'

At this moment, Iran Air 655 was talking to Bandar Abbas, passing its position and estimated 11.15hrs arrival time at Dubai. The Airbus was now 40 miles from *Vincennes*, climbing through 4,000ft at just over 300kt. Bandar Abbas acknowledged, just as the second warning went out from the cruiser, this time on the civil frequency. Meanwhile, the *Vincennes'* identification supervisor was checking through the commercial air timetable to see if any civil flight was due down Amber 59. He found none, maybe because Iran Air 655 had taken off 27min late.

On *Vincennes'* beam, the P-3 Orion continued manoeuvring in a manner which suggested that it was passing targeting information to another aircraft. Coincidentally, whoever was flying the Iranian Airbus realised that he had allowed his aircraft to wander three or four miles off centreline towards the western edge of Amber 59. He corrected that by veering back east toward the centreline; in so doing, he turned the airliner in the direction of the *Vincennes*. This was laid out for all to see on the large screen display in the CIC.

At this stage, the approaching aircraft was tagged as squawking the same Mode 2 code as that of the F-14s which had recently moved to Bandar Abbas. F-14s were air defenders not maritime attackers, and to any knowledgeable operator, even if a Tomcat had been launched to sink the *Vincennes* on a suicide mission, it ought to have been flying much faster and lower than the aircraft spotted by the Aegis system. But time was fast

Opposite: *The route flown by Iran Air Flight 655. The Airbus was flight-planned to Dubai along Amber 59, an airway 20nm wide until it narrowed to 10nm when it split into east and west legs.*

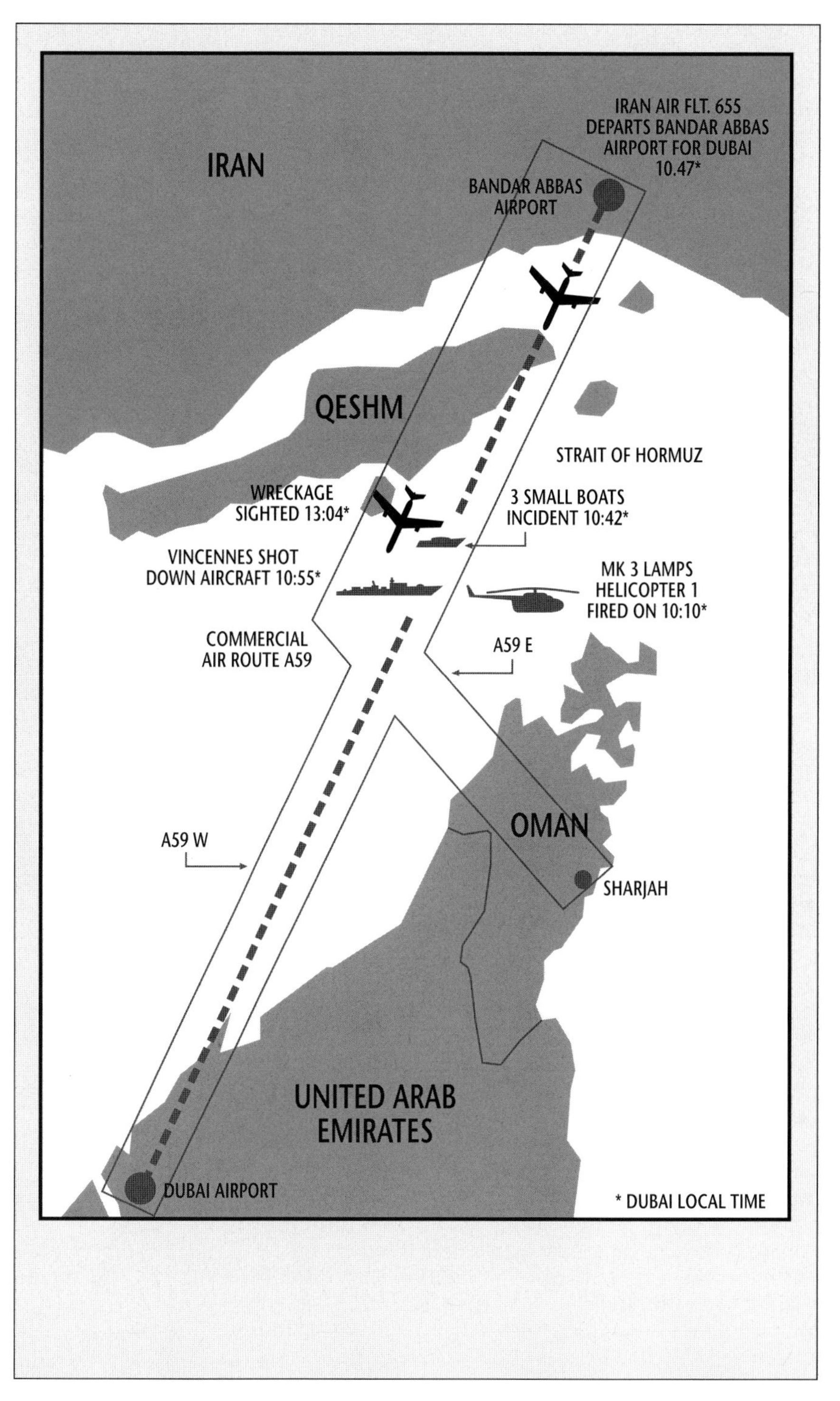
IRAN AIR FLT. 655
DEPARTS BANDAR ABBAS
AIRPORT FOR DUBAI
10.47*
IRAN
BANDAR ABBAS
AIRPORT
QESHM
STRAIT OF HORMUZ
WRECKAGE
SIGHTED 13:04*
3 SMALL BOATS
INCIDENT 10:42*
VINCENNES SHOT
DOWN AIRCRAFT 10:55*
MK 3 LAMPS
HELICOPTER 1
FIRED ON 10:10*
COMMERCIAL
AIR ROUTE A59
A59 E
OMAN
A59 W
SHARJAH
UNITED ARAB
EMIRATES
DUBAI AIRPORT
* DUBAI LOCAL TIME

running out for such refined judgement calls. Although the SPY radar had Flight 655 in its sights for 7min 5sec, Capt Rogers was aware of its possible intent for only 3min 40sec. The surface engagement — 'the wolf closest to the sled' as the subsequent report put it — left very little time for him personally to verify information fed to him by the CIC team. And, unfortunately, that information was seriously wanting.

The Aegis system was designed primarily for fleet air defence in a dense, high speed but broad arena such as the Norwegian Sea; in the open ocean, a possible hostile aircraft can be tracked over hundreds of miles. Combat in the Persian Gulf, which is only about 25 miles wide at the narrowest point of the Strait of Hormuz, is much more constraining. It was subsequently admitted that the amount of time the Aegis system had for verifying threats 'can be seriously eroded when operating close to a land-based airfield'. A captain who hesitated too long while trying to identify an opponent conclusively could lose his ship. That had already happened in May 1987 when the frigate USS *Stark* was hit by two Iraqi missiles launched in error. The *Stark* was severely damaged, 37 crewmen were killed, and Capt Glenn Brindel found his status altered from naval front-runner to real estate salesman almost overnight for failing to take the right defensive action.

The rules of engagement were strengthened after the *Stark* incident such that an aircraft could be fired upon if it came within 20 miles of a US warship and ignored repeated instructions to identify itself. At 10.51hrs on 3 July, *Vincennes'* anti-air warfare co-ordinator was watching the blip 28 miles out as Iran Air 655 was about to climb from 9,000 to 11,000ft. All of a sudden a key figure in the CIC, the tactical information co-ordinator (TIC), asserted that TN4131 had begun descending. As the Airbus came ever nearer, the TIC became ever more insistent, repeating that the closing air target was descending at 1,000ft/mile whenever he could break in on the command intercom circuit; at one point he was 'shouting and yelling'.

With his own life, plus those of 357 crew members and the survival of his ship at stake, Capt Rogers could wait only so long to take action: his STANDARD SM-2 Block II surface-to-air missiles had a minimum range of six miles. Under intense pressure, Rogers had his team transmit seven warning messages to the Airbus — three on 121.5MHz and four on 243MHz — while the nearby frigate *John H. Sides* chipped in with five more. Capt Rezaian did not answer, although he had been chattering away to the Bandar Abbas control tower throughout his brief flight. His last transmission at 10.54hrs was, 'I am at Level one-two-zero, climbing to one-four-zero.' The controller at Bandar Abbas, who was about to turn Flight 655 over to Tehran Centre, replied, 'Goodbye, have a nice flight.'

Vincennes' data banks agreed that the Airbus 12 miles away was passing FL120 but in the CIC, the height being called off by the TIC to the anti-air co-ordinator was '7,000ft, 6,000ft, 5,000ft' as if a kamikaze was running in. Permission to engage having been obtained from Rear Admiral Less, Commander US Joint Task Force Middle East, Capt Rogers turned the firing key at 10.54:22, launching one then another STANDARD missile when the Airbus was 10nm away. Each STANDARD SM-2 carried by *Vincennes*

Opposite: *USS Vincennes, the third US Navy 'Ticonderoga' class Aegis cruiser, firing a STANDARD surface-to-air missile on exercise. When faced with a high-explosive-tipped STANDARD clocking-up Mach 2.5, the Iranian Airbus stood no chance.*

had twice the range and a much enhanced guidance and electronic countermeasures resistance capability than the SM-1 version, but such refinements were totally unnecessary in the circumstances. An airliner cruising along stood no chance, and the first missile homed in inexorably to hit Iran Air 655 at 13,500ft, some 7nm away, 17sec after leaving its rail. On the bridge of the *Montgomery*, the commanding officer watched the flash of the missiles' impact, and the burning wreckage fall. Floating debris and victims were later found in the water some five miles east of the island of Henqam, with nearly 200 bodies subsequently being recovered.

The loss of 290 souls — 278 passengers and 12 crew — from seven different nations on the eve of US Independence Day made the loss of Iran Air 655 the world's sixth worst air disaster to date. In consequence, Rear Admiral William Fogerty, Navy Director of Policy and Plans, US Central Command, headed an investigation team which included a psychiatrist and a physiologist among its members. They worked quickly enough to release their findings on 19 August. The key question to be answered was why a 177ft-long Airbus, going about its lawful and regularly scheduled business along a recognised airway in conventional fashion — at FL140 and around 300kt — should be mistaken for a 62ft-long Tomcat fighter dropping down on a fast suicide-attack mission?

At its simplest, the accident was caused by misinterpretations. For instance, among the wonderful pieces of kit aboard *Vincennes* was the SLQ-32 sensor which detected radar emissions from any attacking aircraft. If it had detected a weather radar signal from the Airbus, that would have helped to identify it as a commercial transport. As it was, the weather at the time — visibility of 8 to 10 miles and scattered clouds at very low level — would not have warranted use of the radar. Capt Rezaian could have navigated visually in those conditions.

Even though Bandar Abbas airport was out of *Vincennes'* line of sight, atmospheric conditions in the Gulf at the time were conducive to a phenomenon known as radar ducting (or bouncing of the signal) which allowed *Vincennes'* transponder to pick up a signal from an F-14 on the ground at Bandar Abbas, possibly taxying out behind Flight 655. Despite the expenditure of $600 million on Aegis, the integration of the cruiser's radar with its IFF system left something to be desired. For example, the IFF operator had to move a range-gate manually to track an incoming target. Unfortunately, the operator in the CIC positioned the range-gate manually on Bandar Abbas and then failed to move it alongside Flight 655's 'hook' as that target climbed away. From that point on, the crew apparently became psychologically disposed to believe that TN4131 was an F-14, irrespective of the fact that they received not a shred of corroboration from *Vincennes'* data bank.

In the crew's defence, all the pieces seemed to fit the offensive picture: the speedboats deployed to distract attention, the targeting P-3, the attacking F-14 not transmitting to avoid alerting the target while sneakily flying along an airline corridor. Thereafter, what psychologists term 'perceptual bias', whereby people see what they expect to see, took over. Accident investigators found that the Aegis air defence system worked perfectly. Crew members simply misread the data offered to them.

The situation was exacerbated when the anti-air warfare co-ordinator — an experienced officer whose job it was to provide air defence advice to Capt Rogers — was reassigned to a fleet control position. His place was taken by a less experienced officer who 'took the information at face value' without checking the character read out on the display screens. In bygone days, old-fashioned raw radar returns could be interpreted by a skilled operator to deduce target size: this was no longer possible with the computer-generated symbols on the *Vincennes'* screens. Much hinged on verbal advice being given out within the CIC as the situation reached its most hectic, and in among the hubbub was the ever-rising voice of the tactical information co-ordinator reporting that the intruder was descending as it got nearer and nearer. Yet the investigators knew, as the *Vincennes'* data bank confirmed, that Flight 655 was then just about reaching top of climb. Their report concluded that 'in the excitement of the impending engagement', the TIC could have misread decreasing range for decreasing altitude, 'or simply misread the character readout' block on the screen in front of him. But the TIC was not alone. Information retrieved from *Vincennes'* data banks 'recorded a flight profile consistent with a normal climb of an Airbus 300', and if the anti-air warfare co-ordinator had made any attempt to confirm the reports he was receiving, quick reference to the console in front of him would have immediately shown that TN4131 was climbing. Instead, he relied on the judgement of two second-class petty officers. Everyone from Capt Rogers downwards could have looked for himself at all of the data including the Mode 3 readout churned up by Aegis. But none of them did. In the final analysis, everyone cocooned below decks in the CIC was expecting to see an Iranian F-14 attacking their ship, so they saw a warplane — and shot it down.

After the misinterpretations came the communication failures. Iranian military traffic controllers kept a close watch on shipping in the Gulf and they must have known about the spat between the *Vincennes'* group and the gunboats. But nobody warned their civilian air traffic colleagues that Iran Air 655's path would take it directly over the developing firefight. Had they known that, 'we would have told the aircraft not to take-off,' said Hossein Pirouzi, airport manager at Bandar Abbas. His staff did not know because the military and civilian controllers at Bandar Abbas did not talk to each other. Air traffic controllers at Dubai, who heard *Vincennes'* warnings, did not think to notify the Airbus. Furthermore, in contravention of UN guidelines for warships operating in areas of commercial air traffic, there was no co-ordination between the US task force in the Gulf and local air traffic control. Communication should have been everyone's business.

What of the repeated warnings transmitted by *Vincennes* and *John H. Sides*? These warnings, which addressed Flight 655 as a fighter, an 'unidentified aircraft' and as an F-14, also called out the aircraft's bearing, speed and height. Only one warning, from *John H. Sides*, quoted the 'squawk' code which the Rezaian crew was likely to recognise. According to a UN International Civil Aviation Organisation (ICAO)

Opposite: *Vincennes' CIC. Hughes PT–525 projection displays presented computer-generated summaries of airborne and surface target data to the CIC team. But in the Persian Gulf on 3 July 1988, the crewmen in Vincennes' CIC became cut off from reality, and that is always a dangerous state to be in when dealing with real aeroplanes.*

investigation, 'The contents of the challenges and warnings issued varied from one transmission to the next. It is uncertain whether the flight crew would have been able to identify their flight and the subject.'

Key evidence, which was reliably given, was that the airliner refused to veer off when warned. We will never know whether the crew understood the warning. Reportedly, the Iranian government had instructed its airline pilots not to reply to such challenges. On the other hand, Capt Rezaian may have become jaded by the frequency of recent US challenges; air traffic controllers in the region had certainly complained about continual American warnings and alerts. Because airline pilots varied in their responsiveness, emergency turns could be a jump from the frying pan into the fire as they could place one airliner on a collision course with another. This actually happened on 8 June when the USS *Halyburton* became edgy at the approach of British Airways Flight 147 from London to Dubai. The warship redirected BA147 right into the path of Balkan Flight 2102 outbound from Sharjah airport, leaving the Dubai controller 'absolutely bloody livid'. Since then, some pilots had got into the habit of ignoring military instructions, and until 3 July they had got away with it.

But in the end, the US Government accepted responsibility to the extent that it agreed to pay *ex gratia* compensation to the relatives of the 290 victims. For the future, it was clear that the misidentification problem could be solved by providing a readout of altitude data directly on to the large CIC screens to aid executives in their decision-making. But there is only so far that you can go down this route. If every screen is filled with every piece of information that could be possibly crucial, senior commanders would be swamped by 'data deluge'. The central question raised by the *Vincennes* incident is whether ever more sophisticated technology is now pushing the fallible humans who operate it beyond their ability to make wise and timely judgements in times of stress when information is often ambiguous or conflicting.

In sum, whatever the gee-whiz technological breakthroughs, there is no getting away from human error, or as some might describe it, human nature. Given such realities of life, there is an obvious peril attendant on missile-equipped warships operating for real in a hair-trigger environment within an area that is 50% covered by commercial airways. Wars beget accidents. It would be terrible if the wit of man, who can design wondrous weapons of war, cannot simultaneously evolve deconflicting rules and procedures that make the terrible loss of Iran Air 655 avoidable in future.

8 The *Challenger* Space Shuttle

Since 1957, when the Soviet Sputnik hurt their pride, Americans have been in love with space. Yet the passion ebbed and flowed. It burnt most fiercely after 1961, when President Kennedy decided to race for the moon, but it started to fade after that great leap for mankind was successfully accomplished. Even before the moon programme, it had occurred to the National Aeronautics and Space Administration (NASA) that it was risky to take on jobs that required going to particular parts of space to do particular things. After putting a man on the moon, NASA flirted with the idea of voyaging to Mars, but for the Agency to thrive it needed to find a task that it could perform more or less permanently.

The most promising answer was to concentrate on capabilities, not destinations, and from this grew NASA's enthusiasm for a permanent, orbiting manned space station with a space truck ferrying people and materials to it. With imagination, the argument went, there could be no running out of interesting things to do on board such a facility. Yet despite spirited lobbying, NASA did not receive permission to build a space station until 1984. Around 1970, not even NASA could escape the federal budget's gravitational pull. With the Vietnam War raging and earthbound poverty to be conquered first, NASA was only able to win support for the truck — a 'Space Transportation System' (STS) that would be able, eventually, to ferry astronauts to and from the space station should one ever be built. The original idea — for a fully reusable space plane fitted out for long stays in space — was pared down to become the Shuttle, and only then on the understanding that it would simultaneously serve military, commercial and scientific needs. The Shuttle had no particular place to go: in effect, it was to be an orbiting laboratory-cum-satellite launch and repair depot. In January 1972, after halving of the original funding, a three-element Shuttle consisting of orbiter, rocket boosters and disposable propellant tank was approved by President Nixon.

About the size of a DC-9 commercial airliner, the Shuttle is essentially a rocket-boosted aircraft embodying rather special technology. The orbiter, which most people think of as the Shuttle, is an aircraft-like vehicle 122ft long with a delta-shaped wing 78ft across, a cargo bay and a tall, vertical fin and rudder with movable flaps to serve as speed brakes. The orbiter is strapped to a gigantic 154ft external tank which feeds its liquid-fuelled engines. The whole assembly is launched vertically into orbit, the liquid-fuel rockets firing in concert with two 149ft tall solid-fuel rocket boosters (SRBs). The boosters are dropped off at altitude and recovered from the sea for future use. The orbiter space plane continues to orbital altitude, where its propellant tank is jettisoned as the engines shut down. The fuel tank breaks up as it drops back through the atmosphere, and is the only unit of the system that is expendable.

USA

In orbit, the vehicle is flown as a spacecraft by means of its orbital manoeuvring system (OMS) engines mounted on the aft fuselage under the delta wing. The space plane re-enters the earth's atmosphere by firing the OMS rockets against the direction of flight, and it descends as a glider. As such, the orbiter weighed in at 150,000lb.

Rockwell International was selected as prime contractor for the orbiter at an estimated cost of $2,600 million over six years. A separate main engine contract went to Rocketdyne, a Rockwell subsidiary. Martin Marietta was signed up to build the huge external propellant tank, Morton Thiokol Chemical Corporation had responsibility for the solid rocket boosters, McDonnell Douglas for the structures and United Space Boosters looked after checkout, assembly, launch and refurbishment. The design process was impaired from the start by what one analyst described as 'continuous penny-pinching'. Problems with the main engines and heat-resistant tiles that protected the craft's thin skin caused persistent delays, and by 1978 the Shuttle was really only being kept alive by the Pentagon's need for more reconnaissance satellites in the midst of Strategic Arms Limitation Talks with the Soviets.

Like most aeronautical ventures, the Shuttle first flew two and a half years behind schedule at 20% over budget. By opting for a reusable orbiter, NASA hoped to be able to push down the cost of doing things in space, but it never managed to cut the cost of routine access to space. The biggest difficulty lay in NASA's inability to meet an over-ambitious launch timetable. Initially, the Agency envisaged a programme of 570 Shuttle flights in the 1980s and 1990s, yet there were only 10 launches during the whole of 1985. The gulf between hope and realisation kept the cost of launching satellites high, due in no small measure to the fact that the Shuttle had been turned into a camel loaded down with so much baggage that it carried nothing efficiently. The decision to have it take on all launch duties for NASA and the Pentagon made the Shuttle far more unwieldy than it might have been. The Pentagon insisted on a much larger vehicle than NASA originally wanted, and much of the extra cost involved adapting the planned propulsion system to carry the greater weight. Once would-be customers discovered that conventional unmanned missiles could launch satellites just as easily, into higher orbit and for less cost, they turned increasingly to Europe's Ariane throwaway rocket despite NASA keeping the price of the more sophisticated but temperamental Shuttle artificially low at $38 million per launch.

The first orbiter was dubbed *Enterprise* at the behest of Star Trek fans. This was really only a test-bed for later models, of which there were four: *Columbia* in 1981, *Challenger* in 1983, *Discovery* in 1984 and *Atlantis* in 1985. These four orbiters were not just satellite launchers. With a 60ft long by 15ft wide cargo bay capable of lifting payloads weighing up to 65,000lb, each orbiter could carry the European Space Agency's Spacelab, a manned orbital workshop with open pallets for scientific instruments. From 12 April 1981 to 28 January 1986, the Shuttle flew 24 consecutive successful missions to form part of a programme that was counted a magnificent success.

STS 51-L *Challenger* was slated for the 25th Shuttle mission. All Shuttle flights were launched to the east from the Kennedy Space Center, on the Atlantic coast of central Florida, using the same two launch complexes — LC-39A

Opposite: *A Space Shuttle being transported out to the launch site.*

and 39B — that Apollo astronauts used when starting their journeys to the moon. Preparation for *Challenger's* tenth journey into space had been painstakingly careful, and for its all-American crew, agonisingly slow. The mission had originally been scheduled to lift off on 20 January from Pad 39B, but the date slipped by five days after the Shuttle *Columbia* ran into delays. Yet when Saturday dawned, *Challenger's* crew learned that a dust storm had developed over an emergency landing ground near Dakar in Senegal, so everything went back another 24 hours.

On Sunday morning, the crew was ready but the weather was not. A cold front was moving down the Florida peninsula, pushing showers ahead of it. While such precipitation does not hinder airliner take-offs, its impact on the Shuttle at the speed it reaches shortly after lift-off could damage the thermal tiles. Once again, every risk had to be minimised, and the fact that *Challenger* could not blast off even into drizzle underlined once again the tight safety margins attendant on human space flight compared to unmanned competitors such as Ariane.

Monday looked much better. For the second time the crew members settled into their couches on the orbiter's two decks, just ahead of the cargo bay. The Shuttle was designed to carry up to seven people into orbit. Two of these — the commander and pilot — were strapped into the flightdeck. On Mission 51-L the commander was Francis R. ('Dick') Scobee, making his second Shuttle trip at the age of

Below: *Mission 51-L astronauts during training. Mission commander Dick Scobee is front right and mission pilot Michael Smith is front left. Mission specialist Judith Resnik is in the central flight engineer position, while Ellison Onizuka is aft left.*

46. The pilot was Cdr Michael J. Smith, a US Navy test pilot and Vietnam veteran aged 40. The remaining seats were filled by technical and scientific personnel, known respectively as mission specialists and payload specialists. Behind Scobee and Smith were Judith Resnik, a 36-year-old electrical engineer, and Ronald McNair, a 35-year-old physicist who was an expert in lasers. On the middeck below were USAF Lt-Col Ellison Onizuka, at 39 both a pilot and an aerospace engineer, Gregory Jarvis, the payload specialist and a Hughes Aircraft electrical engineer aged 41, and in between them, Christa McAuliffe.

Christa McAuliffe, 37-year-old mother of two, personified NASA's efforts to win continued backing for its efforts by blending technology with sophisticated public manipulation. The 'Civilian in Space' programme was designed to capture the public imagination, and it was Ronald Reagan on the presidential campaign trail who decreed that a teacher would be the first to go up. Mrs McAuliffe, a social studies teacher from Concord, New Hampshire, was picked from over 11,000 applicants because she was sufficiently poised and confident in front of the cameras to capture the hearts and minds of the millions of children who would watch her two lessons from space the day after *Challenger* lifted off.

Lying on their backs, the crew of seven could see a bright blue sky ahead of them. The countdown reached T (for take-off) minus nine minutes, and there it stayed for four hours. A sticky bolt was preventing the removal of an exterior-hatch handle, and when Lockheed technicians called for a special drill, its battery was found to be dead on arrival. After 90min of fiddling, an ordinary hacksaw was used to free the bolt. By then, gusts of up to 35mph were sweeping across the Kennedy Space Center; once again, NASA took the prudent course and authorised another delay.

That night, temperatures fell to an unseasonal -3°C though the wind speed fell to 9mph. On Tuesday, 28 January, the morning sky dawned clear and blue. Even before *Challenger's* crew crossed the access arm to strap in for the third time, NASA's ice team had inspected the Shuttle and its towering gantry. They decided that there was no danger of any icicles breaking away on lift-off and harming the heat-shield tiles. Just 20min before scheduled lift-off, they made another check. A Rockwell engineer in California, watching by closed-circuit TV, telephoned to urge a delay because of the ice. But Kennedy Space Center Director Richard Smith, having been advised that there was little risk, allowed the countdown to continue.

'We're at nine minutes and counting.'

NASA Commentator Hugh Harris's voice was relayed over the public address system, and shivering reporters, school children and other spectators cheered. The countdown was past the point where it stopped the day before. On *Challenger*'s flightdeck, roughly the size of that on a Boeing 747, Scobee and Smith ran down their elaborate checklists. The orbiter's main computer, supported by four back-ups, continuously scanned all the data from some 2,000 sensors. They would shut down the entire system in milliseconds if they found anything wrong; nothing was.

'T minus eight minutes and counting.'

Thousands of motorists along that part of the Florida coast pulled off the roads and faced the ocean. Onizuka, Jarvis and McAuliffe had nothing to do but wait like passengers on a $1.2 billion airliner.

'T minus seven minutes, 30 seconds and counting.'

The walkway was pulled away. It could be repositioned within 15sec but the seven occupants were effectively wedded to 143,351gal of liquid oxygen and 385,265gal of liquid hydrogen. Two lines connected the fuel to the orbiter, where they would be mixed at controlled levels to power the spacecraft's engines. The gleaming white SRBs, each packed with over 1.1 million lb of solid fuel, sat stolidly by. Once ignited, they would burn until their fuel was spent, by which time they should have powered the orbiter to about Mach 4 and 140,000ft.

'T minus six minutes and counting.'

At this stage, pilot Mike Smith was given the order to pre-start the auxiliary power units that would operate *Challenger's* control surfaces and swivel its engine nozzles. The last oxygen was pumped into the external tanks.

'T minus four minutes and counting.'

Mission Control reminded the flight crew to close the visors on their helmets.

'T minus three minutes and 30 seconds.'

By now, the Shuttle was operating totally on its own internal electrical power system.

'T minus two minutes, 20 seconds. No unexpected errors reported.'

From then on, Harris's announcements grew into a crescendo as everything continued to look good.

'Ninety seconds and counting. The 51-L mission ready to go.'

After many delays, the *Challenger* crew's waiting seemed to be at an end.

'T minus 45 seconds and counting.'

The launch platform was about to be flooded by powerful streams of water gushing from six 7ft-diameter pipes; this was designed to damp the lift-off sound levels, otherwise the acoustic roar from the Shuttle's three engines could damage the craft's underside. The main engines firing sequence was now under computer control.

'T minus ten...nine...eight...seven...six... We have main engine start.'

The onboard computer could still have aborted everything but there was no sign of any glitch.

'Four...three...two...one... And lift-off. Lift-off of the 25th Space Shuttle mission. And it has cleared the tower.'

At launch, Mission 51-L weighed over 4.52 million lb, the largest Shuttle mission ever launched. At lift-off at 11.38hrs, *Challenger's* two Morton Thiokol solid boosters were each generating nearly 3 million lb of thrust while the three Rocketdyne liquid oxygen/hydrogen main engines each provided 375,000lb of thrust at a 100% throttle setting. It was the most critical and most dangerous part of any space mission — 'When you have that much power,' said Flight Director Jay Greene at the Johnson Space Center, Houston, 'you have to respect it.' No one was flying the vehicle manually at this time; it was all fully automatic while the crew members were being jammed back into their couches by three times the force of gravity.

At 6.5sec into the flight, the main engines automatically throttled up to 104%. Pad 39B's flame pit could not have coped with that amount of thrust at ignition, but it helped *Challenger* clear the pad's fixed service structure.

'Houston, we have roll programme,' declared Commander Scobee as control of the flight shifted from Kennedy Launch Control to Johnson Space Center Mission Control. The flight was only 16sec old.

'Roger, roll *Challenger*,' acknowledged Mission Control's Richard Covey as if he was just any professional air traffic controller. At this stage, the solid rocket nozzles were automatically steered to initiate a 90° right roll. Like a fly clinging to a caterpillar, the Shuttle rolled gracefully on to its back as the fuel tank and boosters assumed the proper down range course for entering orbit. At the same time, *Challenger* began an automatic pitchover of about 20° to place the crew towards a heads-down position relative to the horizon for the ascent.

Below: *Blast off. Note the massive fuel tank and solid rocket boosters to which the orbiter is attached.*

Shortly after completion of the roll, *Challenger's* computer automatically throttled its main engines first down to 94% thrust, then at 35sec, down to 65% power as the orbiter passed through the zone of high turbulence. At this stage the Johnson Center public commentator, Steve Nesbitt, announced that the 'three engines were running normally... Velocity 2,275ft/sec (1,538mph). Altitude 4.3nm. Down range distance (from Pad 39B) 3nm.'

'*Challenger*, go with throttle up,' said Mission Controller Covey after 52sec of flight. This was not an order; it meant that the engines had automatically reached full power and systems were go. The Shuttle was about to endure the most stressful physical forces in the ascent.

'Roger, go with throttle up,' confirmed Scobee at 70sec. It was the last transmission from the spacecraft. NASA's long-range TV cameras had been following *Challenger's* shiny white rocket plume set against a cloudless, dark blue sky. But after the graceful roll the cameras caught an ominously unfamiliar sight — a small orange flickering glow around the lower third of the Martin Marietta external

fuel tank and the right solid-fuel booster. Milliseconds later, this was followed by a much more intense flash fire towards the external tank's hydrogen section (the external tank is actually two tanks, containing liquid oxygen in the upper section and over twice as much liquid hydrogen in the larger, bottom part). Suddenly, there was only a fireball. Piercing shades of orange and red burst out of a billowing white cloud, engulfing the spacecraft as it disintegrated at 47,000ft.

Snaking wildly out of control, the two SRBs flew out of the conflagration clearly intact. They veered widely apart, making a configuration that resembled a giant monster in the sky, its two claws reaching frantically forward. The rocket boosters continued on their curved trajectory until destroyed by range safety officers about 30sec later.

Back in Houston, public commentator Nesbitt kept his eyes on the programmed flight data displayed before him, as yet unaware of the images of disaster appearing on TV screens around the world. There was then a 40sec pause as the horror of it all sunk in. Nesbitt's unemotional tone did not change. Communications with the craft had been severed. He continued, 'We have no downlink.'

Challenger's main tank exploded with the force of a small nuclear bomb, but some investigators believed that the crew members may have remained conscious and aware of the crisis as their module fell for 3-4min before hitting the sea. The horror of it all was a salutary shock for the American people whose brightest and best had soared into space 55 times over 25 years, and whose safe return had come to be taken for granted. In addition to the loss of seven lives — the first US astronauts to be lost in flight — *Challenger* was valued at about $1.5 billion and its payload at over $117 million.

America was numbed by the disaster, and the impact of the tragedy was made even more poignant by the presence on board of Christa McAuliffe. She was the first space coach passenger, not counting a pair of congressional junketeers, and she was inaugurating a people's express to the stars. But it would be wrong to single out one person. The crew of *Challenger* 51-L seemed to represent Everybody's America. One was black, one was Jewish, one was Asian-American from the youngest state, Hawaii, and two were women. In more ways than one, an American dream went down with *Challenger* that day. Therefore it became even more important to discover what went wrong so that the dream could be resurrected and put back on course.

If NASA thought that it would be left to investigate the accident for itself, it was mistaken. Such was public interest in the disaster that President Reagan set up a Presidential Commission on the Space Shuttle *Challenger* Accident with sole responsibility for investigating *Challenger's* destruction. The commission, chaired by William P. Rogers, Secretary of State in the Nixon Administration, included 12 other distinguished people including Brig Chuck Yeager, the first man to break the sound barrier.

The first step, as in any accident investigation, was to gather every scrap of evidence on what went wrong. In this, the NTSB investigators called in for their airliner-crash expertise were fortunate. Unlike the average airliner

Opposite: *Challenger explodes at 47,000ft 73sec after lift-off. The solid rocket boosters were torn from the external tank in the process, and exhaust from the boosters described a 'Y' against the dark blue sky.*

accident, there was much to go on. Everyone knew where the Shuttle went down, and a fleet of 13 vessels, four aircraft and nine helicopters began searching an area that eventually grew to 6,000sq miles of Atlantic coastal waters. In the process, they picked up thousands of pounds of wreckage, including a large section of the orbiter's fuselage and the nose of a booster rocket.

Then there were all those pictures that the whole nation had seen, over and over again; experts would now study them in slow motion with computer enhancements, repeatedly. NASA not only had 80 of its own cameras filming the launch but it also impounded all the film in the 90 remote-control cameras that various news organisations had installed near the launch pad. Finally, there were the billions of signals sent between the doomed Shuttle and NASA computers at Launch and Mission Controls before and during the 73sec of flight.

NTSB engineers laid out recovered debris in a Kennedy Space Center hangar, and they were soon surprised at the amount of heat and flame damage apparently caused at the aft right-hand side of the orbiter even before the explosion. Digital engineering data plus photography of the accident built up a detailed chronology of what occurred on 28 January following lift-off from Pad 39B.

At first, it seemed that the two SRBs were functioning normally. Twenty seconds into the flight, power reduced to 94% and, at about 50sec, the propellant in both solid motors started gradually to build thrust back up again. Sensors on the left motor showed it beginning to increase thrust as planned, but sensors on the right motor showed it was not increasing thrust properly; this indicated that the case had ruptured and that hot, high velocity gas was beginning to escape from the 149ft-long stage.

At about 59sec into the flight, the Kennedy tracking camera captured the leak as a 4ft by 8ft bright plume emerging from the attachment joint between the two aft segments of the right motor. Photography showed this plume growing for about 15sec until it became at least 40ft long, playing like a blowtorch on to the external tank. Chamber pressure from the motor continued to lag until, at around 70sec, it was some 4% below the desired level, confirming that a massive leak was affecting the booster stage.

A second later — just before the explosion — the right motor was providing over 100,000lb less thrust than its left counterpart. The flight control system noticed this difference and began moving both the main engine rocket nozzles and the solid motor nozzles to steer the vehicle and compensate for the lagging booster, which by then had a massive plume spraying from the ruptured joint.

The SRBs were attached to the Shuttle's external tank by struts at their forward and aft ends. The 3,000°C plume from the booster leak was directed towards the booster's lower tank attachment point. In the last second before all data was lost, the aft attachment points for the right booster were either severed by the fire or broken structurally by the abnormal stresses created by the high-velocity leak.

Coincidentally, the booster's plume or structural loads also severed the 17in diameter line which carried liquid oxygen down the outside of the external tank to the orbiter's main engines. A split second after the aft SRB

Opposite: *Solid rocket booster debris recovered from the sea bed. The 11ft by 20ft piece comes from the area where the right aft field joint ruptured.*

attachment points failed, data from the rate gyro systems in that booster showed the bottom of the solid rocket beginning to flip outward, pivoting around its upper attachment point but still secured to the portion of the external tank that separated the oxygen and hydrogen sections. This had the effect of driving the upper portion of the booster into the side of the external tanks, rupturing the oxygen and hydrogen section. Hydrogen was seen to spill out at 73.137sec after launch, followed immediately by a flash between the orbiter and the external tank, an intense white flash at the booster forward attachment point, and then increased flash intensity. The engines, sensing their upper temperature limits, shut down, but to no avail. It was all over in a flash created by the powerful explosion that engulfed *Challenger* at T+73.213sec.

When it came to locating the cause of the disaster, it did not need the brain of a rocket scientist to focus on the right SRB. The Shuttle's SRBs were designed and built by Morton Thiokol Space Division of Brigham City, Utah, while Marshall Space Flight Center, Huntsville, Alabama, was the NASA agency responsible for overseeing SRB operations for the Shuttle programme. Morton Thiokol used conservative specification margins and proven technology to minimise the chances of failure of the motor case, its joints or the propellant itself. Each SRB was put together, segment on top of segment, in the Kennedy Space Center vehicle assembly building. The four segments were mated by tang and clevis joints. Each joint was held together by bands and 177 steel pins, and had two O-ring seals packed with an asbestos-filled zinc chromate putty; the whole was designed to prevent the escape of the 3,000°C gases under 1,000psi pressure that triggered the explosion of Mission 51-L. The inherent strength of motor and joints, despite the leak between the booster's aft and lower centre sections, was what held the stage together for nearly 15sec before the rupture triggered the explosion of the entire Shuttle.

Yet within three weeks of the accident, the investigators determined that the lower assembly joint in the right SRB failed immediately upon ignition on the launch pad — not 59sec later in the climb out. NASA engineering camera views showed a puff of black smoke blowing out of the aft joint of the right booster segment 0.6sec after the motor fired, and black smoke indicated that grease, putty and rubber seals in the joint were burning. The eventual outcome of this burn-through was the emergence of a large plume of white hot exhaust gases from the side of the motor 59sec into flight. 'It seems to me,' said investigating commission chairman William Rogers in early March, 'that the joint (between the aft segment) is the number one villain.'

This diagnosis was confirmed on 13 April by the recovery from the ocean floor of a 4,000lb piece of the aft part of *Challenger's* right SRB: the 11ft by 20ft piece had a large rupture in the aft joint of the booster. So what went wrong with something which up to then had been regarded 'as reliable as hell'? The answer lay in the fact that the original specification had taken insufficient account of unusually cold launch conditions. Temperatures were below freezing the night before 28 January, and when NASA's ice team had surveyed Pan 39B for hazardous ice, their infra-red temperature sensor showed temperatures down around -14°C on the aft right booster. Ground temperature was only 2°C at the time of launch. Unfortunately, the O-ring rubber seals, designed to prevent exhaust leaks between the solid motor segments, lost their resilience in cold temperatures.

Enhanced photographic data showed that smoke vented from the motor at ignition on Pad 39B in a series of closely spaced puffs. These occurred at a rate of three times per second, the same frequency at which the entire Shuttle vehicle resonated at lift-off. Marshall Space Flight Center believed that the motor joint was designed to account for this dynamic motion, but investigating board members found otherwise. They believed it possible that after the motor initially leaked on the launch pad, the leak was plugged temporarily by glassy aluminium oxide — a byproduct of motor combustion. When flame re-emerged 58sec into flight, it was just after the vehicle had been shaken by wind shear as it was flying through maximum aerodynamic pressure. The shaking of the Shuttle at this point flexed the joint and reinitiated the leak. Sod's Law ensured that, after enduring a very unseasonable launch temperature — the lowest temperature at which a previous launch was made was 12°C in January 1985 — Mission 51-L suffered more dynamic motion at this stage than had been encountered on any previous launch.

On the face it, therefore, the loss of Shuttle Mission 51-L seemed to be down to damned bad luck. But not so. A caucus of Morton Thiokol engineers knew exactly what the effect of abnormally low temperatures would have on their booster rocket seals. Between 16.00hrs on 27 January and 08.00hrs on launch day, there were no fewer than 15 separate meetings or telephone conversations betwixt and between Utah and representatives at Kennedy and Marshall Center concerning the effects of low temperatures on booster seals. Morton Thiokol witnesses, led by Allan McDonald, manager of the solid rocket motor project, stated they wanted the mission delayed until temperatures rose to at least 10°C. But McDonald said that opinion at Marshall 'was appalled' at the recommendation not to launch. The ball was firmly thrown back at Morton Thiokol by implying that the mission would go ahead unless they proved that their hardware was not flightworthy. It is hard not to conclude that after eight days of slippage, the need to maintain faltering public interest in NASA's activities overrode a proud, safety-first tradition.

The Rogers Commission found that NASA discovered a booster O-ring design problem as early as 1977 and that Morton Thiokol began treating it as an acceptable risk, even though some Marshall management thought a redesign was essential. While criticising the Marshall Space Flight Center for its booster decisions, the commission also criticised safety oversight, excessive worker overtime and booster test procedures used at Kennedy Space Center under the direction of Marshall and Morton Thiokol. It was because of such wide-ranging and longstanding management deficiencies that the commission called Mission 51-L 'an accident rooted in history'. Consequently, the greatest change to come out of the accident was a massive revamping of Shuttle safety organisation and management.

On 13 June 1986, President Reagan directed NASA to implement commission recommendations for an improved solid rocket motor. The field joint metal parts, internal case insulation and seals were redesigned, the major change to the field joint being the addition of a captive feature that would prevent deflection between the sealing surfaces caused by motor pressure at ignition or structural bending by high winds. No putty would be used in the new design, and a heating system was added to protect the seals from the effects of low temperature in winter. NASA reported that detailed inspection of Mission 26 *Discovery* (the first to fly after the accident) SRBs showed no sign of O-ring seal leakage or erosion after the booster cases were recovered from the sea.

The loss of *Challenger* highlighted three major flight safety deficiencies. First, there was the failure to act. Morton Thiokol engineers and an element of NASA management, principally at Marshall, knew at least five years earlier that the booster joints had serious design problems. They failed to act to halt the flight programme to assess the critical safety implications of the joint design.

The need to avoid costly redesigns was symptomatic of a regime that was both penny-pinching and could brook no delays, which led to the failure to listen. Commission members described as 'beyond belief' the failure of Marshall officials and senior Morton Thiokol managers to heed the warnings of lower Thiokol engineers during the night of 27 January not to launch *Challenger* the next morning because of low temperature effects on the booster seals. Pressure applies to all facets of aviation, and it can often take more courage to say 'No' than 'Go'. Human error engendered by political imperatives make a potentially lethal combination.

Finally, in setting up Christa McAuliffe for the ultimate field trip, NASA fell victim to its own publicity. Space flight in 1986 was never as routine as a bus journey, but it had to be made so to counter cooling public interest in space plus a declining NASA budget. NASA had to mount increasingly glamorous missions to secure the funding it needed to launch more glamorous missions; in effect it had become a prisoner of itself, surviving principally in order to survive. Whatever the rights or wrongs of such an environment, there should have been no place within it for sightseers. In forgetting that, and in placing a civilian on top of a Roman candle, NASA transformed an American dream into a nightmare.

9 Deep Freeze

Ice is an insidious killer of aeroplanes. It can build up quietly but relentlessly on wing, tail or control surfaces to the extent where it seriously impairs aerodynamic lift or control through changed aerofoil shape or weight. Furthermore, a layer of ice only 0.5in thick over the inlet lip of an intake can 'flame out' an engine. All this has been well known for years, and major airports which frequently experience icing conditions are equipped with de-icing gear that is capable of removing or preventing the accretion of ice on even the largest airliners. Even so, ice is a tricky customer that brooks no half-hearted response.

The first fortnight of January 1982 provided an exceptionally harsh introduction to the new year in Washington, DC. Snow battered the northeast USA continually and 13 January was no exception; Washington National airport had to be closed for more than an hour in the early afternoon to enable fallen snow to be removed. Flight operations resumed shortly before 15.00hrs. For those passengers booked to fly Air Florida Flight 90 to Fort Lauderdale and Tampa, the prospect of warm, sunny skies just a couple of hours away must have seemed particularly inviting.

Boeing 737-200 (N62AF) was allocated to Flight 90. After N62AF was de-iced, and was tentatively moved back on the slippery ramp, the airliner took its place in line among numerous other aircraft waiting for departure clearance. Unfortunately, as the Washington National had been closed for over an hour, everybody was pressing for take-off clearance while in-bound aircraft held in the stack were clamouring even louder to be allowed to get down on the ground. Flight 90 finally received clearance to take off on Runway 36 nearly an hour after the de-icing had been completed. And during that hour, the 737 had sat under a continuous light-to-moderate snowfall with the temperature remaining below freezing.

Aircraft behave like any other machines in these conditions. Their skin cools to below freezing, and if that

Above: *Ice on a helicopter blade.*

skin then comes into contact with water droplets from snow, rain, drizzle or fog, ice builds up just as on a car when it is left outside overnight. By the time Flight 90 received clearance to go, considerable snow and ice had once more accumulated on it; there was up to 0.5in thickness on the wings.

Even a small amount of snow will disturb the smooth flow of air over a wing. This will cause airflow separation at a lower angle of attack than normal, which increases the stalling speed while reducing lift. None of this was welcome on a fully laden 737 and, just as critically, getting airborne safely was not helped by simultaneous build-up of ice in the compressor inlets of the two Pratt & Whitney JT8D turbofan engines. As N62AF began its take-off roll, the CVR conversation between Capt Larry Wheaton and First Officer Roger Pettit, who was flying the aircraft from the right-hand seat, showed that things were not as they should have been:

Captain: 'Holler if you need the wipers.'
Captain: 'It's spooled.'
Unidentified: 'Really cold here.'
First Officer: 'Got 'em?'
Captain: 'Real cold.'
First Officer: 'God, look at that thing.'
First Officer: 'That doesn't seem right, does it?'
First Officer: 'Ah, that's not right.'
Captain: 'Yes it is, there's eighty (kt).'
First Officer: 'Naw, I don't think that's right.'
First Officer: 'Ah, maybe it is.'
Captain: 'Hundred and twenty.'
First Officer: 'I don't know.'

The flying pilot recognised early on that the thrust lever settings, engine instrument readings and progress in the ground roll did not 'seem right'. After noting the 80kt airspeed indicator reading, Capt Wheaton did not respond to further comments from his first officer; he seemed to remain focused on the airspeed indicator while Roger Pettit continued the take-off roll.

Besides degrading performance, the accumulation of snow and ice caused the 737 to pitch up immediately after it got airborne, activating its stick-shaker stall warning device. Given the reduced power output and degraded lift capability, N62AF stalled and fell out of the sky. The 737 had reached between 200-300ft before it started to descend, turning slightly to the left in the process but generally maintaining a northerly course. This took Flight 90 across the Potomac River and straight towards the parallel 14th Street road and railway bridges. It is believed that the crew first lowered the nose and then raised it to try and maintain height, and in the final moments they strove to increase power, but all to no avail. Roger Pettit was heard to say, 'Larry, we're going down, Larry!' and the captain replied, 'I know it.' In the circumstances, there was not much more to say.

It was then 4 o'clock in the afternoon. The streets of downtown Washington were clogged with traffic as offices closed early on the worst day of winter. The 14th Street

Opposite: *Washington National airport and, just to the north of it, the 14th Street road and rail bridges over the Potomac that connect central Washington with Virginia.*

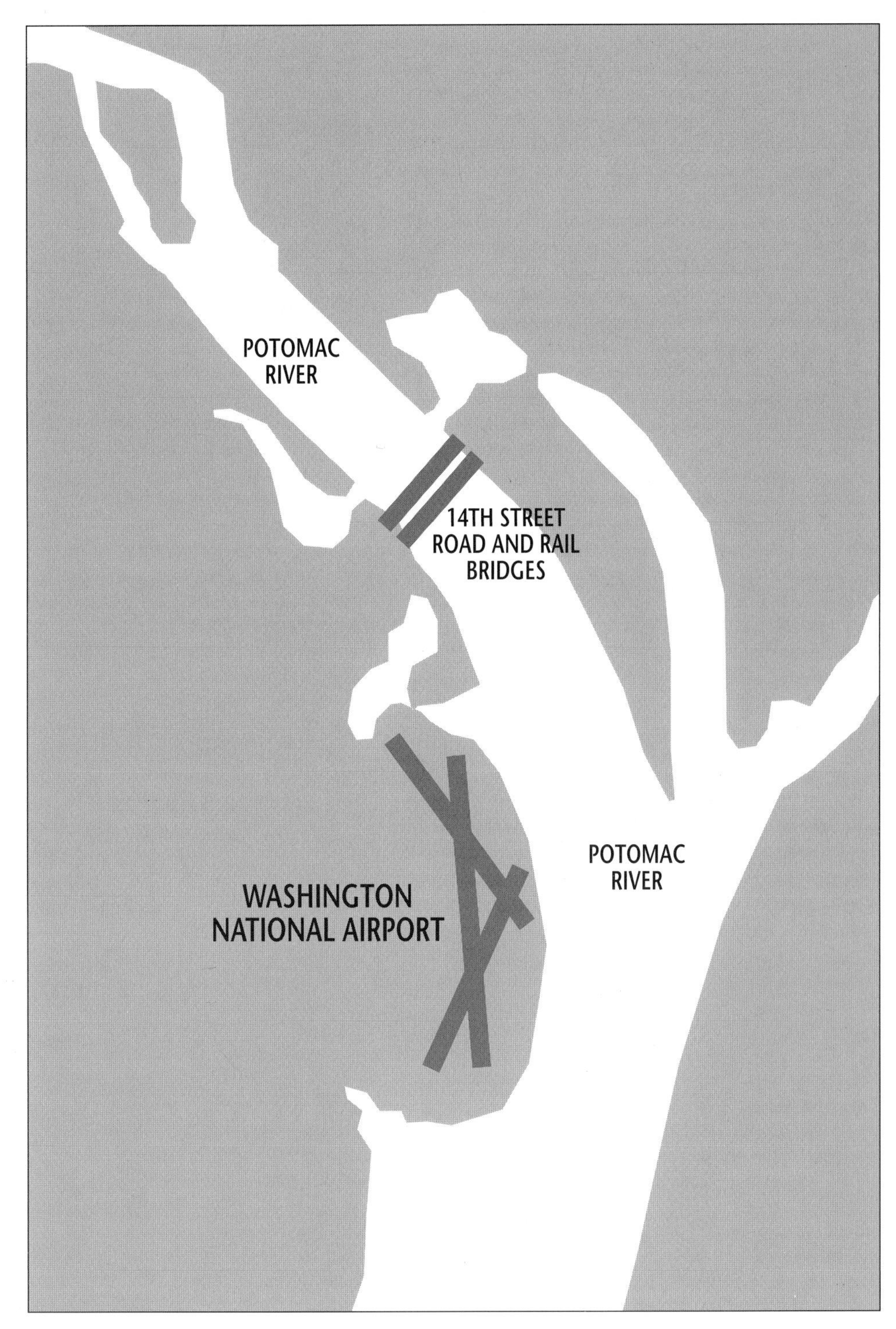
POTOMAC
RIVER
14TH STREET
ROAD AND RAIL
BRIDGES
POTOMAC
RIVER
WASHINGTON
NATIONAL AIRPORT

bridge, leading to commuter-land in Virginia, was especially jammed solid as Flight 90 hove into view. Its undercarriage was still down and the flaps partially extended when N62AF, flying nose-up at an angle of between 30-40° and wings level, struck the northbound span of the 14th Street bridge about one mile from the up-wind end of Runway 36. Skimming over the roadway, the airliner's tail skid hit the bridge parapet and a car, slicing through both like a knife. The rear fuselage and tailplane destroyed six occupied cars and a truck, and then tore away a section of the bridge, before the rear empennage broke off. The fuselage and wings carried on beyond the bridge before plunging down into the ice-covered Potomac River. All but five of the 79 people on board Flight 90 were killed, including four crew members, together with four occupants of the vehicles. The four passengers and one flight attendant who survived were all seated at the back of the rear fuselage which remained partially above water. They all suffered injuries — Joseph Stiley survived despite suffering two broken legs — as did four people on the bridge.

A primary factor in the crash was the decision by the crew to take off despite knowing that there had been a build-up of snow on the 737's wings. The pilots may have been influenced by the prolonged delay before departure. Had they opted for a second snow and ice clearance session, it would have necessitated a return to the ramp and another long delay pending arrival of, and attention from, an overstretched de-icing team; this would not have been received with universal acclaim by the fare-paying passengers.

Above:
The empennage of Air Florida's Boeing 737 being recovered from the Potomac.

But there were three other areas where the flight crew could have helped rather than hindered. The first was in the use of reverse thrust to back the 737 away from the terminal. Heat from the engines and thrust reversers, when combined with the blowing snow and slush, could have deposited a wet mixture on the airframe and especially the wing leading edges, which subsequently froze. There were also indications that the pilot-in-control intentionally parked the 737 close behind another aircraft, hoping to use its heat efflux to remove the snow from N62AF's wings. In the event, the heat may simply have turned the snow, which would otherwise have been blown off during the take-off roll, into a slushy mixture which then froze on the leading edges or turbofan inlet nose cones.

The second factor which directly contributed to the disaster was the decision to continue the take-off run despite the instrument indications and the gut feeling that everything did not 'seem right'. Since the first officer was doing the flying, Capt Wheaton should have been the more attentive to the overall signs, but it was Roger Pettit who seemed to be most observant. The final NTSB report made particular mention of 'the captain's failure to reject the take-off during the early stages when his attention was called to anomalous engine instrument readings'. The flight crew may have been rushed by an air traffic plea for 'no delay on departure' because of landing traffic; margins were being pushed to their limits to the extent that an Eastern Airlines 727 touched down on Runway 36 even before Flight 90 had lifted off. But the actions of the pilots, and particularly those of the captain in failing to make a decision to abort in the 20-odd seconds he had before the 737 passed its Stop Speed, probably reflected a general lack of experience in cold weather operations. It was believed that he missed meaningful exposure to the harsh winter climate of the Eastern US due to the rapid expansion of Air Florida in the late 1970s and early 1980s, wherein pilots were upgraded faster than the industry norm to meet the demands of expanding schedules.

Among other contributing factors was the long delay between de-icing and take-off clearance. Ramp personnel also used an inappropriate glycol de-icing solution mixed for -2°C, when the actual temperature was -4°C, due to the non-availability of the appropriate monitoring device. But the third and most crucial crew oversight was the simple failure to turn on the engine anti-icing system. While ice was building up in the engine compressor inlets, it was also blocking the engine discharge probes. These probes determined correct thrust setting, and post-accident tests confirmed that blockage of these tubes resulted in a higher thrust reading than the engine was actually delivering. That was the ultimate explanation why the take-off did not 'seem right' to the first officer; the fully-opened engines were pushing out much lower thrust than normal, he could sense that acceleration was not as it should have been, yet the instrument indications gave no cause for concern.

Had the powerplant probes not been blocked, the correct readings would have been indicated and the thrust correctly set. The ice would still have been on the wings, but the thrust should have been sufficient to get N62AF out of trouble. On the other hand, had the icing been so severe as to remain in the tubes notwithstanding the use of anti-icing, the first officer would have been unable to set the power at the correct reading, and this would undoubtedly have prompted the flight crew to discontinue the take-off. It was for this reason that the NTSB accident report ascribed the crew's failure to use the engine anti-icing system as the direct cause of the disaster.

It would be unjust to paint too damning a picture of human error or neglect because the crash itself brought forth some stirring individual cases of bravery and self-sacrifice. In cold, wet conditions where hypothermia can kill within minutes, the survivors, who were able to cling to the wreckage, owe their lives primarily to the crew of a US Park Police Long Ranger helicopter who arrived on the scene to hoist or tow them to safety. Two bystanders actually jumped into the freezing Potomac to help; one of them, Lenny Skutnik, saved a woman who lost her grip on the rescue line. Another passenger who lived through the crash unselfishly passed the line to other survivors; by the time the helicopter returned for him, he had slipped beneath the surface and drowned.

But such human courage must not obscure the fact that this accident should never have happened in the first place. In the rush to make up for lost time, a whole host of people from de-icers to air traffickers cut corners in the perfectly understandable quest to return some semblance of order to flight schedules that had been severely disrupted by foul weather. Yet it is in just such circumstances that Sod's Law comes into play. Sod's Law says that if something can go wrong or awry, it will do so at the most inappropriate time. I remember when we could not clear an airfield after a particularly heavy overnight snowfall because the snowplough was under the drifts somewhere and no one could remember where it had last been parked. Similarly, when the weather was foulest on that fateful day in January 1982, the de-icing fluid monitor was not available to the de-icers of Flight 90. And in among those making haste at the expense of speed were the flight crew, who committed the ultimate sin of forgetting to switch on the engine anti-icing! Life is like that. When you have all the time in the world, you never forget the anti-icing; it is when you are up to your armpits in alligators that you have to take the greatest care in not getting bitten.

Although pre-flight de-icing is costly in terms of material and man-hours, these penalties have to be accepted as no other safe options are open to airline or pilot. Poor Capt Wheaton was probably deluded into believing that the take-off was progressing normally until it was too late to become safely airborne or discontinue the take-off. Experience can be worth more than all the dials and instruments put together. That is why the human brain must remain on the flightdeck — the final safety link is, 'Trust Your Gut Feeling'.

Yet safe operation in icing conditions is much more than simply taking precautions on the ground. Icing seriously degrades performance in the air, and as icing is usually associated with other aircraft-threatening weather phenomena, the combination can be daunting as was shown in February 1980 over Massachusetts, USA.

Nearly four decades earlier in February 1943, the British Under-Secretary of State for Air made public a report outlining the types of British commercial aircraft that an official committee considered would be required after the war. The committee, chaired by the distinguished aviation pioneer Lord Brabazon of Tara, originally recommended five types that ran the gamut from a 100-ton London-New York non-stop express airliner to a small feeder transport for internal services. In between was a four-engined aircraft

Opposite: *Britannia G-BRAC in 'Redair' livery at Cranfield while appearing in the BBC TV series 'Buccaneer'. G-BRAC, in the colours of Redcoat Air Cargo, crashed at Billerica, Massachusetts, on 16 February 1980.*

which eventually metamorphosed as the Bristol Britannia. Despite being the world's first long-range turboprop airliner, only 82 of these Proteus-engined 'Whispering Giants' were eventually built because, by the time the Britannia entered public service in 1957, it was clear that the future lay with jets such as the de Havilland Comet and Boeing 707.

Of these 82 Britannias, 20 were finished to Model 253 standard for use as strategic transports by the RAF. The Britannia retired from RAF service after the 1974 defence cuts, whereupon some moved into civilian life where the Britannia offered much as a charter aircraft plying at high load factors. One ex-RAF Britannia was given the civilian registration G-BRAC in June 1978, and it was in the colours of Redcoat Air Cargo Ltd that G-BRAC arrived at Boston Logan airport, Massachusetts, from Belize in Central America at 15.10hrs on 15 February 1980. Known as Flight RY103, G-BRAC uplifted 33,435lb of freight in readiness for a trip to Shannon, Ireland, the next day.

At about 11.00hrs on 16 February, a Flight 103 crew member entered the National Weather Service office and requested a 500mb prognosis chart for the North Atlantic. He returned a few minutes later accompanied by other crew members. They requested forecasts for several airports in the British Isles and the weather briefer remembered that the crew appeared to be in a hurry. After some toing and froing, the crew finally departed; the briefer subsequently remembered that he had neglected to tell them about a forecast of occasional severe icing in precipitation over New England.

At 11.55hrs, the crew went to file a flight plan while the flight engineer and ground engineer prepared the Britannia for departure. Since a snowfall during the night had left considerable deposits on the aircraft, the flight engineer requested that local ground service sweep the snow off and apply de-icing fluid to G-BRAC.

The flight engineer and ground engineer watched the snow being swept off before leaving the area while de-icing was performed. Subsequently, the flight engineer recalled that snow fell intermittently during de-icing and

before engine start, and that the snow was wet. He stated that he walked round the aircraft after de-icing was completed, and he found the control surfaces to be properly cleared. However, this could not be corroborated; no ground witnesses saw the flight engineer or any other crew member check control surfaces after de-icing was completed.

The remainder of the flight crew arrived at G-BRAC about 10-20min afterwards. Once the crew of six plus two passengers were aboard, the de-icing crew gave the wings and horizontal stabiliser a 'fast shot' of de-icing fluid. After the four 3,960hp Bristol Siddeley Proteus engines roared into life, the Britannia remained parked for 20-25min with them throttled back to idle. At 13.51hrs, Flight 103 was cleared to Shannon via flight plan route.

The aircraft taxied from the ramp at 13.55hrs. The ramp supervisor saw snow and possibly frost beginning to accumulate on the wing leading edges as G-BRAC left the area, but the flight engineer stated that the entry guide vane heat was on before taxying and he recalled seeing the outside air temperature at 6-8°C. He remembered the snow being 'wet, it was mild. The snow that we were getting was very, very wet snow, very wet. Each time we stopped, I leaped out of my seat, peered through the radio window and there was no build-up of snow or ice on the leading edge of the nacelle, around the intake or the leading edge of the mainplane... I am convinced that there was no appreciable ice build-up on the aircraft before we started to take off.'

At 14.07:11hrs, Flight 103 was cleared for take-off from Runway 33L and to 'turn left on to 315° at the two mile DME fix after departure'. At 14.08:41, the tower controller asked if 103 was rolling yet, and at 14.09hrs the brakes came off with the first officer doing the take-off. Two watching snowplough drivers thought that everything seemed normal. The Britannia got airborne after some 7,000ft of take-off roll. The flight engineer noted that the aircraft encountered severe turbulence — which he likened to a 'high frequency buffeting' — immediately after lift-off, and that the turbulence remained constant during the climb. The CVR transcript confirmed that the normal after-take-off checklist items were completed including gear and flaps up. Max continuous power was set at 14.10:20hrs and the first officer called, 'Two DME, going left.' At 14.11:36hrs departure control advised Flight 103, 'Low altitude alert, check your altitude, climb and maintain 9,000.'

Flight 103 replied, 'We're passing 1,200ft, cleared to 9,000.' The CVR then revealed an intra-cockpit comment, 'Bloody rough, isn't it?' At 14.12:07 the captain asked, 'Got the de-icing on?' The flight engineer replied, 'Affirmative'.

At 14.12:49hrs the departure controller passed a second low altitude alert: 'RY103, low altitude alert. Check your altitude immediately. Shows 1,400ft; the minimum safe altitude in that area is 1,700ft.' The captain replied, '103 roger, we're getting a lot of chop here.'

At 14.13:03hrs the first officer said, 'Cowl heat and icing can go off now, can't it?' The flight engineer replied, 'Cowl heat's not on'. Although the flight engineer could not note any airspeeds being flown, he did recall that the first officer raised the nose after the second low altitude alert when the captain directed, 'Go at V_2 plus three then, Jack'.

Departure control asked 103 to turn right on to 360°. The captain replied, 'We're getting some pretty severe down-draughts here'. The controller responded with the reassuring news that the air was a bit smoother above

4,500ft, but that was rather academic for a Britannia crew that found itself going down rather than up. The CVR recording was illuminating (figures in brackets indicate time):

First Officer: 'Full power.' (14.14:08)
Flight Engineer: 'I think we'll have a little bit more power out of it.' (14.14:14)
First Officer: 'Yes.' (14.14:17)
First Officer: 'Bloody thing's going down.' (14.14:26)
First Officer: 'Any icing?' (14.14:30)
Navigator: 'No, there's nothing on the wings.'
First Officer: 'Going down.'

At 14.14:35 the captain called departure control to ask, 'Are we close to high ground here? We just don't seem to be climbing.' The controller asked if Flight 103 wished to return to Boston. Transmissions on the CVR began to break up at this stage but the following internal cockpit conversation was recorded after the captain confirmed that they were flying IFR:

First Officer: 'Okay, do you want to jettison, Bill?' (14.15:22)
Captain: 'Yeah, start jettisoning fuel.' (14.15:23)
First Officer: 'You take control now.' (14.15:25)
Captain: 'Okay, my stick.'
Navigator: 'You're very low, I can see the ground.' (14.15:35)
Flight Engineer: 'Yeah, we're dumping fuel.' (14.15:38)
Captain: 'Get round here you bugger.' (14.15:57)
Captain: 'Controls are frozen.'
First Officer: 'Get some power up.' (14.16:00)
Flight Engineer: 'We have full power now.' (14.16:02)
Captain: 'We're in a stall.' (14.16:05)
Captain: 'Look out.' (14.16:07)
Captain: 'Hold on.'
Sound of impact. (14.16:08)

Witnesses on the ground reported seeing the aircraft below the clouds on occasions, and entering or leaving cloud at different locations along the route. They said the speed appeared to be slow, the aircraft nose-high, the engines at full power and the wings 'wobbling'. G-BRAC was in a climbing attitude but not gaining altitude; eventually it passed through trees and struck the ground on a heading of 050°. The Britannia came to rest adjacent to an industrial area and just short of a housing estate in Billerica, about 16 miles north-northwest of Boston Logan and about eight minutes after take-off from Runway 33L. A severe post-crash fire erupted immediately. Five crew members and the two passengers perished in the crash and subsequent fire; only the flight engineer survived, albeit seriously injured.

The subsequent investigation, carried out by the NTSB because this British-registered aircraft had perished on US soil, found the crew to have been properly certified with no medical problems and G-BRAC to have been maintained in accordance with regulations. Based on the evidence, the investigators considered several possible causal areas — power loss, airframe or flight control malfunctions, weight and balance, crew member actions, and meteorological conditions including wind shear, turbulence, down-draughts and icing.

Of these, the engines were found to have been developing full power at impact, and if engine icing occurred it was not sufficient to cause the aircraft to descend and crash. Nor, despite the CVR reference to 'controls frozen', did the evidence support a control problem; the pilot would not have been able to increase the angle of attack during the descent as was demonstrated by performance analysis. The Safety Board found aircraft weight and balance to have been within limits. This left meteorological conditions and crew member actions.

The weather in the Boston area during the morning and early afternoon of 16 February was characterised by low overcast and obscured skies with visibilities ranging from a half to two miles in snow and fog. Temperatures were slightly below freezing with winds from the east at 7-14kt. From the evidence available, it was clear that the type of precipitation and temperatures aloft varied widely within a relatively small area. Nevertheless, the NTSB believed that intermittent areas of moderate to severe icing existed. Analysis also showed that Flight 103 probably encountered 29-33kt of wind shear above 1,000ft.

Given the 45-60min delay between de-icing and take-off, during which it was snowing intermittently and the surface temperature was near freezing, the NTSB concluded that ice and snow accumulations on the Britannia's lifting surfaces combined with slush on the runway to produce a longer than normal take-off roll. Ice and snow accumulations were also the major factor in the lower than predicted initial climb. Performance analysis revealed that drag remained fairly constant throughout the climb to 1,700ft, though the board was at a loss to explain why neither pilot mentioned the poor climb rate. The flight engineer's description of the turbulence immediately after take-off as 'high frequency buffeting' suggested that part of the aircraft's wing was stalled. Yet, while these conditions would have decreased the climb capability of the aircraft, they were not sufficient to explain the total loss of performance.

When the departure controller issued the second low altitude alert, the Britannia was climbing and, given the lack of high terrain ahead, it would eventually have ascended to altitude and transited to Shannon. Analysis showed that the aircraft began to lose additional climb performance around the time of the second alert. The crew's only comment was, 'We're getting some chop here' before the captain said, 'Go at V_2 plus three'. Two reasons probably prompted this instruction. The second low altitude alert may have led the captain to suspect that G-BRAC was approaching high terrain, and it was pretty certain that the airliner encountered down-draught with associated wind shear at about 1,600ft. Suspecting that the severe down-draught was the cause of their problems, the captain instructed the first officer to fly at an airspeed that would give a better climb gradient; the first officer raised the nose to fly at 136kt (V_2 plus 3kt).

Under most conditions 136kt would have given a better climb gradient, but with the airframe icing conditions that probably existed, the increased angle of attack did nothing.

Opposite: *Part of the wreckage of a BOAC Britannia which crashed in thick fog outside the village of Winkton, Hants, on Christmas Eve 1958. The size of the fin gives some idea why the Britannia was known as the 'Whispering Giant'.*

In fact, the existence of airframe icing put this speed below optimum climb performance and it would have accelerated the accumulation of more ice, further depreciating performance. Once G-BRAC was in that situation, heavy ice accumulations occurred in a very short time such that lift capability was degraded to the point where flight was no longer possible. The flight engineer did not use wing heat throughout the flight because he did not want to bleed off any precious thrust. The lack of wing heating probably made no difference; the resultant loss of torque might easily have compounded the already deteriorating situation. From the minute the first pilot set 'V_2 plus three', the descent of Flight 103 in a nearly stalled condition was inevitable.

The Safety Board was emphatic that the accident was not caused by wind shear, turbulence or down-draughts. These were merely factors which

degraded the climb capability to a point where the low altitude alerts were issued and airspeed was bled off to gain height. The overwhelming factor was the pre-existing and rapidly accumulating airframe ice. Recovery could have been accomplished from almost any situation other than that facing G-BRAC on that afternoon. It is possible that if the airliner had not encountered moderate to severe in-flight icing, it would have continued to climb safely. Conversely, if the aircraft had not departed with pre-existing ice or snow on the airframe, it might have been able to overcome the in-flight icing conditions. What the 'Whispering Giant' could not do was overcome them both. Therefore, these two factors were considered to be the cause of the degraded aerodynamic performance.

The NTSB noted that the failure of the flight crew to obtain an adequate pre-flight weather briefing plus the failure of the National Weather Service to advise the crew of severe icing conditions were contributing factors. Perhaps the crew, having flown in from sunny Belize, were still mentally attuned to Central America rather than freezing New England. But it did not take undue perception to see the actual weather conditions at Boston Logan, or to conclude that time and effort would be well spent in obtaining accurate *en route* weather forecasts.

The crew of Flight 103 ought to have made themselves aware of the environmental conditions, and their possibly hazardous effect on aircraft performance, because it was too late to catch up with reality once they got airborne. The flight engineer stated that he saw no build-up of snow or ice before take-off, but no one could see the entire wing from his position or from any other part of the cockpit — and we are not talking about icebergs here. Wind tunnel tests showed that even small amounts of wing surface roughness, including ice, snow or frost, could seriously degrade lift capability.

In the final analysis, if the crew of Flight 103 had really understood the ice threat to their aircraft, and had known that there was no high ground in the way of their climb-out, they would have survived. Safe airmanship is all about knowing what you are up against. In icing conditions as much as in any bad weather, it never pays to fly blind. And this advice applies as much today as in 1980.

Below: *A British Midland Airways Viscount 813 at Southend in 1972. G-OHOT, the Viscount 813 which crashed near Uttoxeter on 25 February 1994, was an identical aircraft with a similar history before being employed by the Southend-based British World Airlines to carry Post Office parcels.*

There are still a large number of perfectly serviceable elderly aeroplanes around, many of them being flown on *ad hoc* contract work for which they are particularly well suited. One of them was Vickers Viscount 813, G-OHOT. This turboprop first flew in 1958, since when it had been modified to the freighter/passenger configuration, and it was employed by British World Airlines on a Post Office 'ParcelForce' flight on 25 February 1994. Using the callsign 'British World (BWL) 4272', G-OHOT was scheduled for a 19.30hrs departure from Edinburgh to Coventry. But, in anticipation of worsening weather *en route*, the Viscount was dispatched 50 minutes early. It was sleeting when both pilots inspected the aircraft surfaces for residual ice and slush, but they found none and there was no need to de-ice the aircraft.

The general *en route* weather was forecast as rain and drizzle with complete cloud cover at various levels between 600 and 16,000ft. The forecast warned of moderate icing in cloud, and severe icing in nimbostratus cloud.

BWL4272 took off at 18.43hrs with the captain as handling pilot. In view of the weather conditions, the ice protection system for all four Rolls-Royce Dart turboprop engines (each delivering 3,025hp) and the airframe was selected to 'on' throughout the flight. The Viscount climbed to FL190 in cloud. At 19.13hrs the crew commented that the de-icing system was working well and, seven minutes later, the first officer remarked that there was a little surface ice on No 4 engine but it was shedding. He also said that there was some ice on the spinner but none on the wings.

At 19.28hrs, Manchester radar cleared BWL4272 to FL150 with a routeing via the Lichfield NDB to Coventry. Four minutes later, as the Viscount was in the descent, No 2 engine failed. Just as the crew was completing its shutdown drills, No 3 engine started to run down. The captain instructed the first officer to 'get an immediate descent' and to 'declare an emergency'. This he did, though without using the pro-words 'Mayday' or 'Pan Pan'. Manchester immediately cleared the aircraft to descend to FL70 and then to FL50 before passing control to Birmingham radar. At this point, the failure of Nos 2 and 3 engines had deprived the aircraft of its only source of wing and tail de-icing. The respective airframe de-icing switches should have been selected 'off' but this checklist item was apparently not performed.

At 19.36hrs, BWL4272 was further cleared to 2,500ft and the first pilot made 'an emergency request' for a diversion to Birmingham, an airfield with which both pilots were familiar. Two minutes later, as G-OHOT descended through FL84, No 4 engine failed. Almost simultaneously a successful attempt was made to restart No 2, but subsequent attempts to restart the starboard engines during the descent met with no success.

While No 4 engine was running down and No 2 was starting up, the pilot flew an inadvertent 170° turn to the right until the aircraft was heading north. The Birmingham controller queried this and suggested an easterly heading for East Midlands airport. East Midlands was nearer and its weather was not unreasonable, but the crew did not take the hint. A short time later, the Viscount made a series of turns; another heading passing by the Birmingham controller for East Midlands went unacknowledged.

Between 19.40hrs and 19.43hrs, G-OHOT turned left on to south. During this time the first officer battled to restart No 3 engine, and the commander stated that he was losing control in yaw and needed his torch to read the instruments. At 19.42:20hrs, the aircraft descended through its cleared

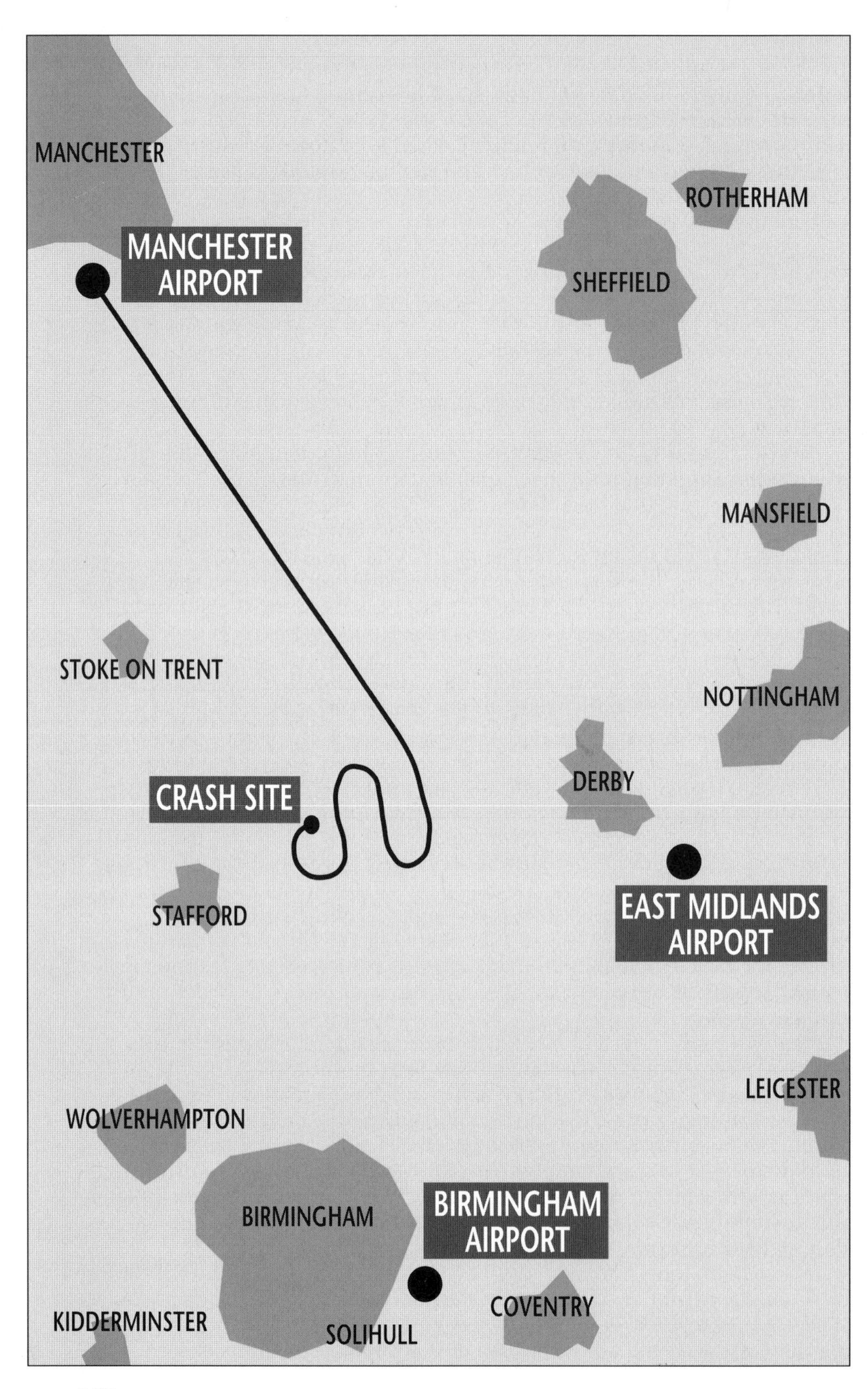
MANCHESTER
MANCHESTER AIRPORT
ROTHERHAM
SHEFFIELD
MANSFIELD
STOKE ON TRENT
NOTTINGHAM
DERBY
CRASH SITE
STAFFORD
EAST MIDLANDS AIRPORT
LEICESTER
WOLVERHAMPTON
BIRMINGHAM
BIRMINGHAM AIRPORT
COVENTRY
KIDDERMINSTER
SOLIHULL

altitude of 2,500ft. It appeared to be unable to maintain altitude, despite the nose being raised a second or two later. This adjustment in pitch, while momentarily arresting the descent, bled 45kt off the airspeed before descent had to be resumed to maintain flying speed. At 19.44hrs the first officer again tried to restart the starboard inner; then the captain exclaimed, 'We're going to stall'.

Between 19.45:17 and 19.45:28hrs, the first officer transmitted two Mayday messages but, probably because of failing power supplies, neither was heard on the ground. At about 19.46hrs, G-OHOT struck a down-sloping part of Drointon Wood, southwest of Uttoxeter, Staffs. The collision with mature trees led to massive disintegration of the aircraft structure which came to rest in a field on the edge of the forest; thereafter, an intense fire consumed the cabin section. The 32-year-old aircraft commander received fatal injuries in the impact, but the seriously injured first officer was freed from the aircraft by two passers-by and moved to safety.

Vickers Viscount G-OHOT had flown over 50,000 hours but AAIB investigators found that both the aircraft and its crew were fit and properly licensed to carry out the flight. It was the extreme nature of the meteorological conditions through which both aircraft and crew had flown that initiated the sequence of engine failures and electrical problems which led to the accident. The vulnerability of turboprop aircraft in conditions of severe airframe and engine icing has long been known, not least because the flight levels at which turboprops operate coincide with the conditions most likely to produce heavy ice accretion. The meteorological forecasts available to the crew, plus the decision to bring forward dispatch time in order to beat the bad weather, should have alerted the pilots to what was ahead once they left Edinburgh. In the event, they appeared to take no action to avoid bad weather. This would have been possible from reference to the Viscount's weather radar which was serviceable, yet it appears that the crew did not even switch the radar on; this was a serious and surprising omission from the safe and effective operation of the aircraft.

The Viscount encountered very heavy icing in thick cloud from the start of the descent — the first officer recalled snow and ice re-forming on the windshield immediately behind each sweep of the wiper blades — and, as the airframe was almost unprotected throughout most of this period, the rate of accretion would have been high. Furthermore, as these conditions persisted almost down to ground level, there was little opportunity for the considerable deposits to melt off.

Following consecutive failure of Nos 2 and 3 engines, cold air would have passed through the heat exchangers and entered the ducting, adding internal icing to that on the external unprotected surfaces. Five minutes later, when No 2 engine was restarted, ice shedding may have occurred on the port wing (provided that the heat exchangers and ducting were not obstructed) but the heating effect was probably diluted in the tail area and non-existent in the starboard wing. This condition, leading to asymmetric drag and lift, persisted for the eight minutes before impact.

The captain's attempt to level the aircraft at the assigned 2,500ft by raising the Viscount's nose cost him 45kt. In order to maintain flying speed thereafter, it was calculated from the radar plot that descent was continued at an

Opposite: *The path taken by BWL4272 on 25 February 1994, showing the changes in direction in its last few minutes.*

airspeed of about 138kt. Even then, directional control was lost and the drag from the unfeathered propellers of the failed starboard engines (blade angle evidence showed that neither was feathered on impact) must have exacerbated the situation. Since the average Viscount would have lost directional control at around 120kt, the difference was explained by the presence of a significant quantity of snow or ice on the fin.

The Rolls-Royce Dart engine began life in 1945, and around 7,000 of these classic turboprops were produced. They have a long record of successful restarts in the air, but G-OHOT's failed engines lacked anti-icing protection. They also remained in almost continuous icing conditions down to ground level, so their intakes were most probably choked by ice and snow when the restarts were attempted; tests carried out in 1957 showed that the rapid ingestion of 3-3.5lb of wet ice was all it took to 'flame out' a Dart. Yet, having regularly practised double engine failures, the crew of BWL4272 should have been able to cope with what was more of an 'abnormal' rather than an 'emergency' situation. But their actions, or lack of them, compounded the situation rather than eased it. For a start, the captain should have been mindful that, having lost Nos 2 and 3 engines, G-OHOT no longer had any airframe icing protection and it needed to be got out of the icing environment as quickly as possible. Then, even during the first engine shutdown drill when there was little pressure on the crew, the emergency checklist shutdown drills were not carried out. Whatever actions the first officer took, he did from memory. Consequently, omissions and errors of procedure were made including failure to reset the fuel trimmers (which reduced the chances of engine restarting), failure to close the airframe de-icing systems (allowing the worst possible airframe icing to occur), omission of electrical emergency actions (resulting in the loss of several electrical services), and failure to feather the propellers of the two failed engines.

There was no lack of endeavour on the flightdeck of BWL4272, but the work was compartmentalised and unfocused rather than sensibly and rationally co-ordinated. At no time did the first officer comment on the incorrect airspeed being flown, while the commander made no observation about the emergency checklist not being used. Similarly, when the first officer informed ATC that an emergency existed, the decision to divert to Birmingham had already been made. The level of activity on the flightdeck was increasing at this stage, but it did not excuse the lack of geographic orientation which prevented the crew from realising that East Midlands was considerably closer and where G-OHOT would almost certainly have been able to land. It was very unfortunate that the crew never asked ATC for 'a diversion to the nearest suitable airfield'.

Notwithstanding the double engine failure, proper emergency drills should have ensured a successful completion of the flight by the four-engined G-OHOT. But this is very easy to say. The crew was up against a rapidly worsening situation at night, in IMC, in severe icing and turbulence, in an aircraft heavily contaminated by ice or snow and with engines failing sporadically. This was way beyond any exercise that may have been practised previously in a nice, warm simulator, and few pilots would have much remaining presence of mind by the time the essential electrics deteriorated, the flightdeck lighting and crew communications went on the blink — the pilots were eventually forced to shout in order to communicate with each other — and the commander lost directional control of his aircraft.

In summary, there were four interrelated factors that led to the loss of Viscount G-OHOT on that freezing February day. The first was multiple engine failures resulting from flight in extreme icing conditions in thick cloud. The Viscount eventually exceeded Civil Airworthiness 'continuous (exposure) maximum' criteria for its type by over 11 minutes in the frontal nimbo-stratus.

The second was incomplete performance of emergency drills by the crew. Besides the build-up of ice or snow in the intakes, the probability of engine restarts was prejudiced by the fuel trimmers remaining at zero, maintenance of an airspeed outside the recommended envelope and diminishing electrical power. Although a solitary No 2 generator was working during the final minutes of the flight, its power was not directed to the essential services because the emergency procedure for the electrical system had not been properly followed.

The third factor was crew actions in securing and restarting the failed engines, which limited the power available. The drag from two unfeathered propellers plus the weight of the heavily iced airframe led to a loss of height and control before the chosen diversion airfield could be reached. And finally, poor crew co-ordination reduced the potential for emergency planning, decision-making and workload sharing. Consequently, the crew had no contingency plan for avoiding the forecast severe icing, nor were they aware of the relative position of a closer diversion airfield which they could have chosen had they made more efficient use of air traffic services.

The failure to think and plan ahead while still safely on the ground in Edinburgh was the first omission. Even when consecutive engine failures deprived G-OHOT of any airframe de-icing, once No 2 engine was restarted the aircraft should have been able to maintain height and continue the flight. But the circumstances required clear thinking and decisive action, and in the words of the AAIB accident report, 'The deteriorating situation escaped this particular crew and they never successfully caught up with it'.

Stress narrows the field of vision. As the emergency progressed on BWL4272, the captain became increasingly preoccupied with flying the aircraft manually in turbulent conditions, while the first officer was left on his own to cope with several emergencies. The CVR revealed hardly any discussion between the two pilots about the options available to them and the best course of action. The captain needed to stand back to obtain an overview of his problems without being immersed in the detailed handling of the situation. With 20/20 hindsight, he should have used the autopilot to free his mind in order to command and interact with his crew.

In stressful situations, individual pilots need help or prompting from other crew members. Knowledge of, and familiarity with, emergency procedures and checklists can often be the only certainty to hang on to when abnormal conditions arise. The latest buzzphrase — crew resource management — is all about working and planning together, which in turn demands that time be set aside for realistic and credible crew training. Crew members are there to monitor and assist each other such that the combination of their efforts exceeds the sum of its parts.

None the less, the most significant factor in the loss of BWL4272, and the consequent failure of the mail to get through, was the encounter with severe icing conditions. It has been suggested that lessons learned the hard way when transport aircraft were predominantly piston-driven have been diluted or lost now that modern jet airliners can transit rapidly through levels where severe icing is likely. But that welcome development should not mask the fact that severe icing conditions are far from rare over the UK. Records show that there is a one-in-six chance of meeting an equally severe weather frontal zone to that which befell BWL4272 in February. To date, most aircrew have used good airmanship to keep out of trouble. And this must remain the key for as long as meteorological conditions exist which can generate a level of ice accretion greater than that which engines can tolerate, and which meteorologists have difficulty in forecasting with unerring accuracy.

Weather is the last great frontier in aviation and mankind is no nearer defeating ice, snow, wind shear or turbulence than when the Wright Brothers first flew. Good airmanship is the only sure defence, and wise passengers should grin and bear it if their pilot decides to delay take-off, or hold off from landing, until dangerous weather has cleared. It is better to travel safely than never to arrive!

10 FIRE ALARM

The first man-carrying aeronautical device got airborne in 1783 by virtue of an iron grid on which straw was burnt. The heat thus generated rose through the open neck of an envelope to warm the air enclosed therein. This lifted the contraption, including a basket built around the iron grid and any brave soul willing to stand inside it. Within two years, an attempt was made to cross the English Channel. This balloon was filled with lighter-than-air hydrogen, and it did not need a genius to predict that the mating of an upper half filled with very inflammable hydrogen and a lower section inflated by fire burning beneath was a rather high risk occupation. So it proved. The vehicle managed to take off from the French coast and rose to about 1,500ft before the hydrogen ignited. The world's first airborne fatality was killed by fire.

Life did not get any easier for some time. These early free-flying machines were at the mercy of the winds until the 1890s when Dr Wolfert fitted a Daimler-Benz 'horseless carriage' engine to an airship. However, the combination of exhaust sparks and hydrogen gas was another potential show-stopper, and the inevitable detonation killed Wolfert and a passenger in June 1897. Since then, fire in the air and its dreadful effects have remained one of the greatest fears facing airmen and women.

There are some memorable airports around the world, and generally the most impressive to fly into and out of are those whose runways are surrounded by water. From my experience, the most mind-concentrating of these is Hong Kong International. Being short of building land, Hong Kong has a long history of filling in shallow sections of its natural harbour. In 1924 permission to reclaim an area to the east of Kowloon City was given to two Chinese businessmen, Messrs Ho Kai and Au Tak. Their intention was to establish a 'garden city', but although reclamation was completed successfully, the scheme foundered through lack of capital. The land, now named Kai Tak after its two creators, reverted to the Hong Kong Government who found it to be most suitable for an airfield capable of accepting all aircraft then in operation, while being adjacent to Kowloon Bay for flying, boat and seaplane operations. An airstrip was opened in 1925, since when Kai Tak has become one the world's major airports. But notwithstanding a great increase in international traffic, Kai Tak was constrained by geography to a single 13/31 runway. This runway was to grow to just under 11,000ft in length by the 1970s, which involved the levelling of hills to the northwest of the airport and the dumping during 1955-58 of over 11 million tons of fill into the sea.

It was out over Kowloon Bay that Trans Meridian Air Cargo Flight KK3751 took off on a non-scheduled cargo service from Hong Kong to the UK, with Bangkok as the first stop, on 2 September 1977. Assigned to the task was G-ATZH, a Canadair CL-44 long-range swing-tail commercial freighter based on

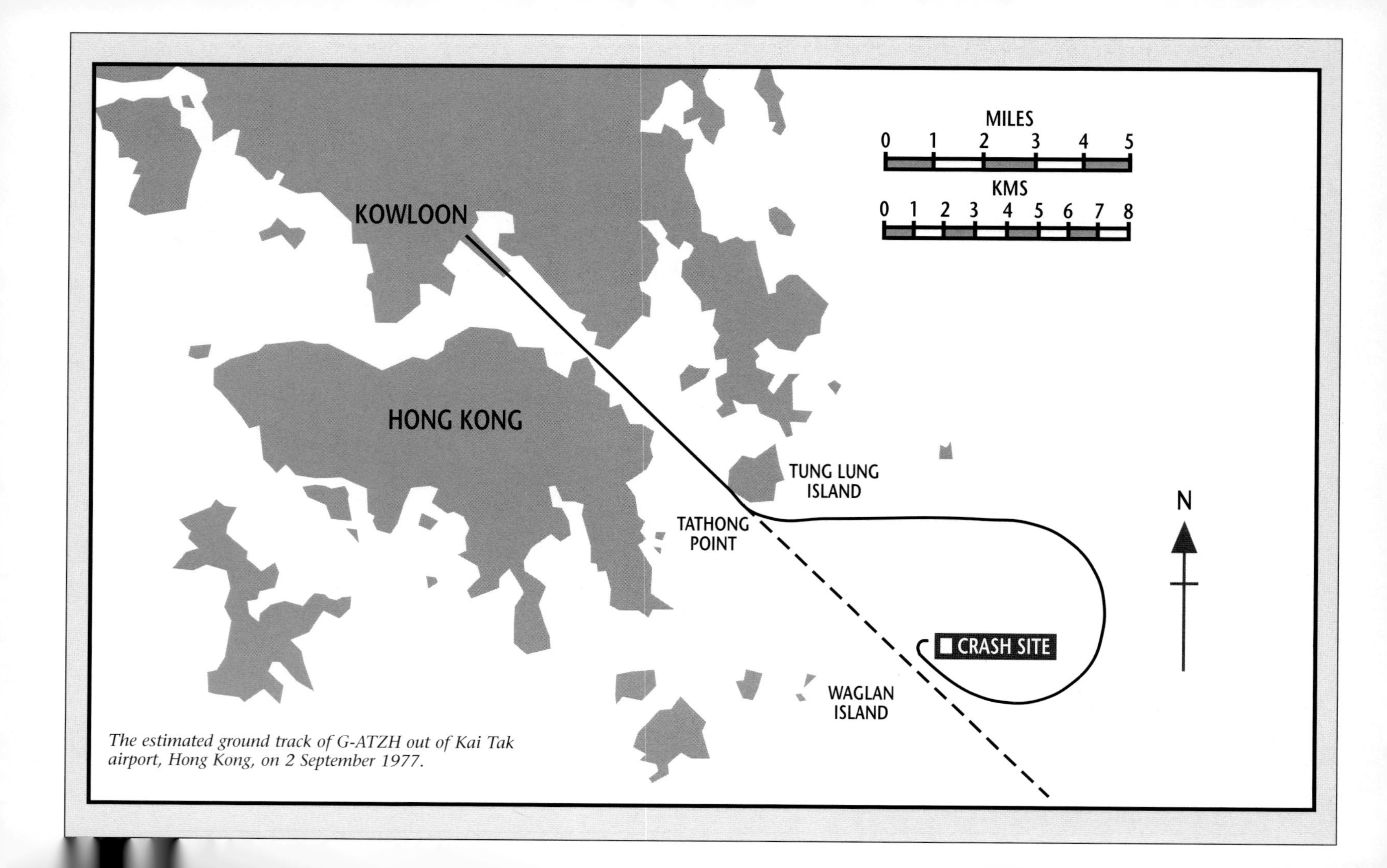

The estimated ground track of G-ATZH out of Kai Tak airport, Hong Kong, on 2 September 1977.

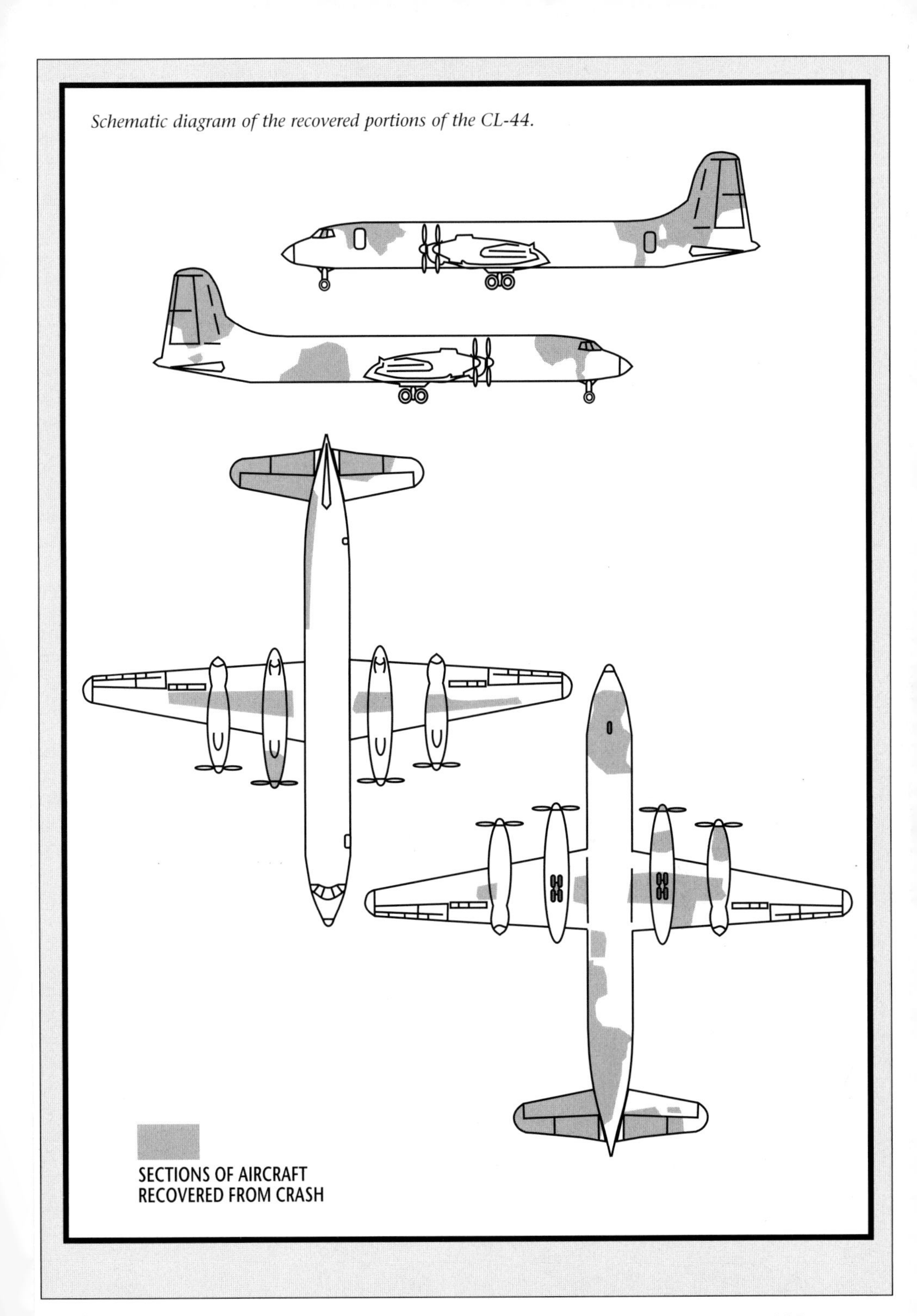

Schematic diagram of the recovered portions of the CL-44.

the Bristol Britannia and powered by four Rolls-Royce Tyne turboprops. The CL-44 with four crew members on board — captain, first officer, navigator and engineer — took off from Kai Tak's Runway 13 at about 00.31hrs. As it became airborne, a puff of smoke was seen from the rear of No 4 engine. According to an eyewitness, this was followed by a torching flame from the jet pipe exhaust which persisted for 10-12sec. The crew was alerted to the smoke emission by ATC, and R/T reports from the freighter indicated that the starboard outer engine had failed and had been shut down. As the CL-44 climbed out on a southeasterly heading, it appeared to trail a thin stream of dark smoke behind the right wing from the vicinity of the engines until it was 2.5nm from the end of the runway at a height of about 450ft above the water. Around one minute after take-off, Tower requested the crew's intentions; the controller was told that these were to 'go out', dump fuel and then return to Kai Tak Airport.

Shortly after 00.33hrs, G-ATZH was cleared to contact approach Control. At 00.33:51 the first officer reported that the aircraft was climbing through 1,000ft and would shortly dump fuel, a process the crew estimated would take 10.5min. The first officer acknowledged the Approach controller's suggestion to take up an easterly heading from the outer marker, away from the most populated areas, and to climb above 4,000ft before jettisoning a mass of kerosene. The aircraft was identified on radar turning east.

At 00.35:28hrs, the first officer informed Approach that the CL-44 would have to return to Hong Kong without dumping fuel because of an engine fire. The aircraft was given radar vectors for an immediate approach to Runway 31, and cleared to descend to 2,000ft on the sea level altimeter setting (QNH) of 1,007mb. According to radar returns, G-ATZH made a right turn and took up a heading of 295°. During the turn, the crew were told that their position was 12nm southeast of the airport and that they were cleared to continue the approach visually if at any time this was possible. The first officer advised that although the flight was in VMC, radar vectoring was still necessary.

As the aircraft closed with the extended centreline at 00.38hrs, the commander warned ATC to have fire vehicles standing by; he advised that the crew had seen flames 'but they seem to be extinguished at this time'. Shortly afterwards he reported, 'There's still smoke coming out of it (the engine).' At 00.38:17, the first officer indicated that control of the aircraft had been lost when he said, 'We're going in'. The last message received from Flight KK3751 before his voice faded out was, 'The engine's come off, the engine's...' In parallel, radar returns faded on the Kai Tak screen at a position 8nm southeast of the 31 threshold. Further attempts to contact the flight crew proved unsuccessful.

Witnesses on a merchant ship approaching Waglan Island saw the aircraft on fire during the final part of its flight. This developed into a fireball, almost wholly engulfing the CL-44 which then made an uncontrolled descent into the sea approximately 2.5nm east of Waglan Island. G-ATZH sank immediately. ATC alerted the emergency services and a major sea and air search was mounted. An area of floating debris consisting of cargo, landing gear, wheels and tyres, fire extinguisher bottles and an inflated life raft, was found approximately 4nm southeast of Tathong Point. But there were no survivors.

Opposite: *Boeing 707 in BOAC livery as Flight 712 would have appeared on 8 April 1968.*

Despite the fact that this accident took place relatively close to one of the world's major cities and that the chartered water depth in the vicinity of where Flight KK3751 went down was around 100ft, only some 25% of the wreckage was recovered even after a protracted salvage operation. The non-recovery of the bulk of the aircraft structure, in particular the No 4 engine and propeller, together with the aircraft's FDR and CVR, typified the obstacles that can be faced by aircraft accident investigators, and made it impossible to determine the cause of the fire on board G-ATZH.

The available evidence indicated that No 4 engine experienced an internal fire resulting in jet pipe and breather pipe overheat. This developed to the extent where the aircraft structure became so weakened by the intense and extensive fuel-fed fire that No 4 engine and the right-hand wing outboard of it fell off. This caused the CL-44 to go out of control.

The decision by the crew, after shutting down No 4 engine, to dump fuel before landing strongly suggested that if an engine fire existed at that time, they were unaware of it. If there had been indications of a persistent engine fire, there would have been a strong incentive to land quickly. Modern aeroplanes are built quite robustly, wings do not burn off in a nanosecond, and therefore the first priority in any emergency situation should be to keep within the parameters of the safe flight envelope. It will give next of kin little satisfaction to know that their nearest and dearest carried out simulator-perfect emergency drills as they spun into the ground. It is therefore most improbable that the crew of Flight KK3751 would have hung around for 10.5min dumping fuel — they would have thrown the CL-44 back on the ground. If in doubt, there must be no doubt. The timeless lesson for all flight crews is never to waste time in the air messing around if there is a hint of fire.

But getting aircraft down quickly is not the be all and end all. On 8 April 1968, British Overseas Airways Corporation (BOAC) Flight 712 was scheduled to depart at 15.15hrs from London Heathrow for Sydney, Australia. The aircraft in question was Boeing 707 G-ARWE, and the crew for the first leg to Zurich was Capt C. W. R. Taylor, First Officer F. B. Kirkland, Second Officer J. C. Hutchinson and Flight Engineer T. C. Hicks. The second officer would normally have monitored operations from the jump seat immediately behind the captain, but Check Captain G. C. Moss sat there on this occasion and so the second officer occupied the navigator's station.

With 116 passengers and six cabin crew in place, the 707 taxied out from Terminal 3 in fine and sunny conditions with almost no wind. G-ARWE lifted off from Runway 27L at 15.27hrs but just after the undercarriage retracted, and the captain was about to call for power reduction in the interests of Heathrow noise abatement, a jolt was felt throughout the airframe accompanied by a distinct bang from the port side. Simultaneously, No 2 engine thrust lever kicked towards the closed position and instrument readings showed that engine to be running down. Capt Taylor called, 'Engine failure drill', and Flight Engineer Hicks began the necessary emergency action.

As he fully closed No 2 thrust lever, the undercarriage warning horn sounded as it should have done. Both the flight engineer and the check captain reached for the horn cancel switch but as Capt Moss got there first and operated it, Hicks instinctively but erroneously pressed the fire bell cancel button. He then reached for the engine fire shut-off handle above the pilots' instrument panel but, on remembering that the captain had called for an engine failure rather than an engine fire drill, he did not pull it.

Seconds later, Capt Moss looked out of the port cockpit window and saw that No 2 engine was on fire in a big way. He told Capt Taylor and advised, 'You'd better turn back and land as quickly as you can'. The fire warning light in No 2 fire shut-off handle was now illuminated and Taylor called, 'Fire drill!' First Officer Kirkland transmitted a Mayday and Flight 712 was cleared to Runway 05R as an alternative to the longer circuit round to land back on 27L.

Meanwhile Flight Engineer Hicks, having begun the engine failure drill, changed to the fire drill; he completed its Phase 1 from memory as required. Opening his own copy of the 707 checklist, Hicks then went through Phase 2 of the drill. On finishing his radio transmissions, First Officer Kirkland tried to assist by reading the fire drill checklist aloud, but Hicks told him it had already been done.

On the other side of the cockpit, Capt Moss continued watching the engine fire through the side window while helping Capt Taylor position the airliner for landing. This was far from easy because G-ARWE was already at 3,000ft and 225kt when the fire broke out, and the turn back on to finals for 05L was tight. Furthermore, there was no approach slope guidance on the relatively short 7,700ft runway.

Ninety seconds after the fire started, and just as Taylor called for undercarriage and full flap to be lowered, the flames weakened the structure of No 2 engine pylon to the point of failure. The 707's magnesium alloy engine pylons were designed to burn through in the event of an uncontrollable fire, allowing the whole engine pod to drop off before damage was done to the primary wing structure. But even after G-ARWE's port inner engine left for pastures new near Egham, the fire continued to burn fiercely. Notwithstanding, Capt Taylor judged the approach well and the 707 touched down smoothly less than 400m in from the runway threshold. As soon as G-ARWE came to rest, there was a massive explosion in the port wing, intensifying the fire and flinging burning debris over the top of the airliner. Before the other three engines could be closed down, Taylor ordered his crew to abandon the

Opposite: *London Heathrow, showing Flight 712's take-off and landing paths and (inset) the approximate circuit flown by G-ARWE during its short time aloft.*

flightdeck while the cabin staff urged passengers out through the overwing hatches. The spread of burning fuel under the fuselage burst the port forward and rear galley escape chutes. Most passengers then had to wait in line to leave through the forward galley door on the starboard side. The exit of passengers was orderly at first but then the rear section of the fuselage was breached by fire. Four passengers in this area, including an elderly woman normally confined to a wheelchair, together with Stewardess B. Harrison who had been marshalling them to the rear galley door chute before it burst, were overcome by heat and smoke. They were the only fatalities, though 38 of the 112 passengers who escaped from the burning 707 sustained injuries of varying severity.

No 2 engine was recovered from a water-filled gravel pit some five miles from the 05 threshold. Its No 5 low pressure compressor turbine had disintegrated due to metal fatigue, and jagged pieces had hurtled off through the engine casing. The fire broke out when one of these high-speed fragments severed the main fuel feed pipe, allowing fuel to gush on to the hot engine under pressure from the booster pumps. These booster pumps worked at the rate of some 225litre/min, which is why turning them off was a crucial item in the 707 fire drill. But that never happened on Flight 712. Post-accident examination of the wrecked G-ARWE showed that none of the four fire shut-off handles (one for each engine) had been pulled, and that neither the fuel booster pump switches for the main wing tanks, nor the switches on the flight engineer's panel for the fuel shut-off valves, had been turned off.

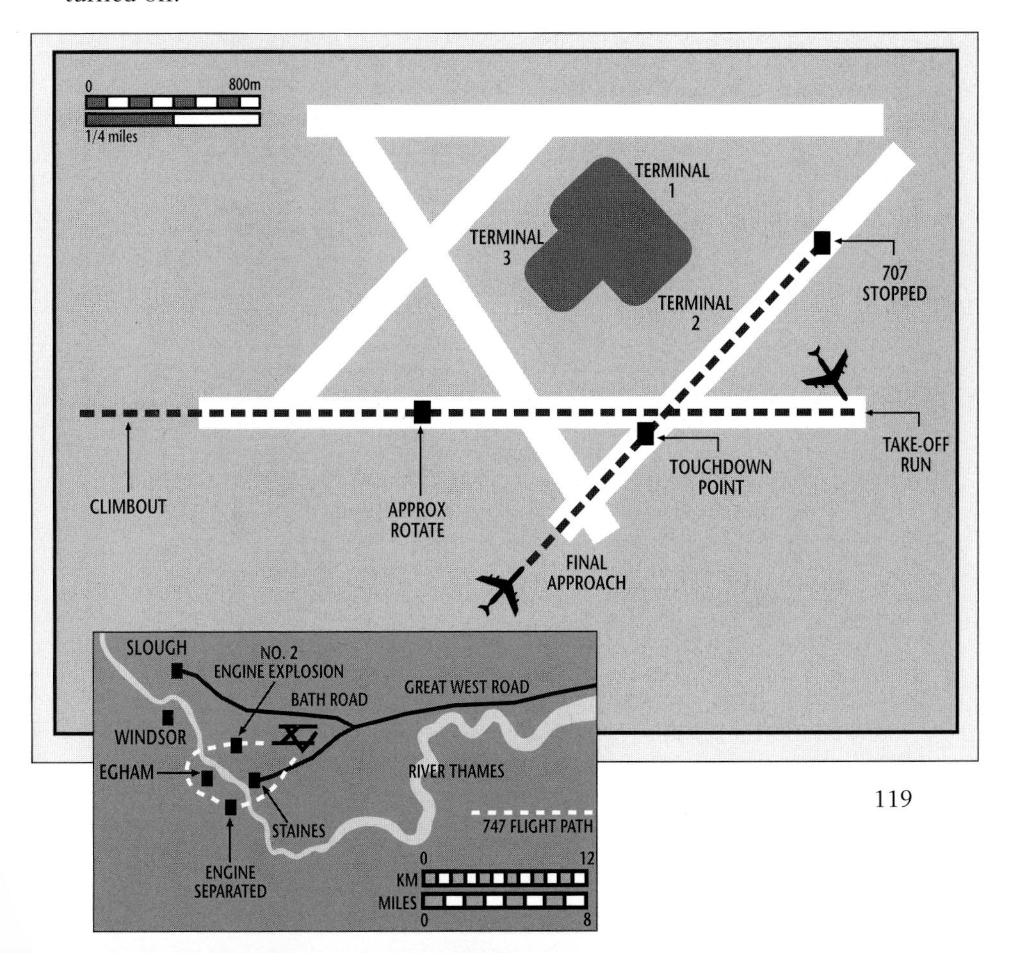

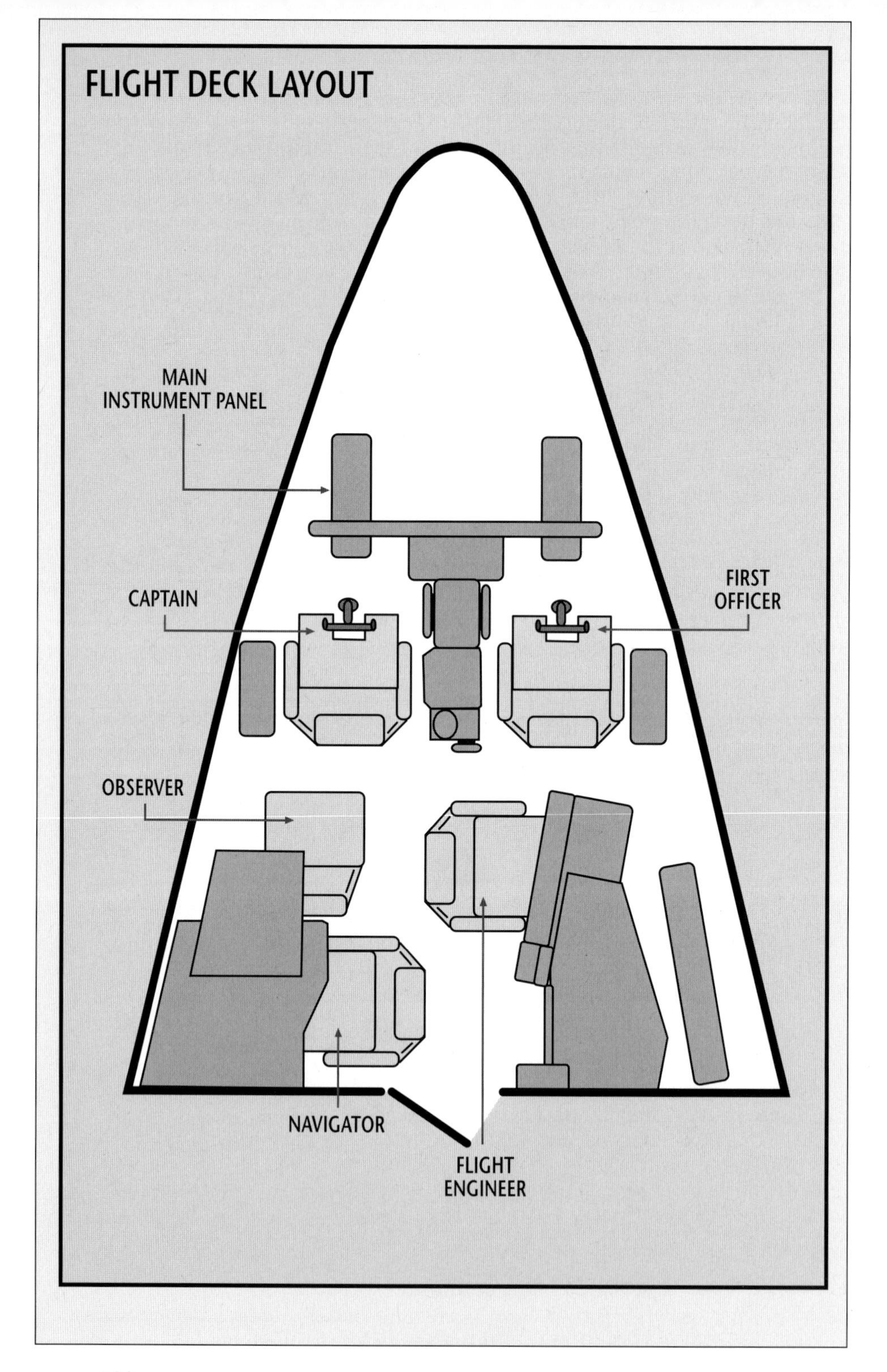
FLIGHT DECK LAYOUT
MAIN INSTRUMENT PANEL
CAPTAIN
FIRST OFFICER
OBSERVER
NAVIGATOR
FLIGHT ENGINEER

Because the fuel shut-off valves remained open, the fire continued to burn fiercely. Once No 2 engine fell from the aircraft, the wing fire carried on burning because it was being fed from fuel flowing from the stub of the severed fuel pipe. Even after the airliner came to a stop on the runway, the fuel pumps continued to run, feeding the fire for another 20sec or so until their electrical wiring was burnt through. But soon after this, the explosion in the port wing released even more fuel, intensifying and spreading the fire.

Airliners such as the Boeing 707 rely on extinguisher systems that use inert gas to smother an engine fire within its cowling. On G-ARWE, this system was controlled through four fire shut-off handles located on top of the pilots' coaming. In the event of an engine fire, a red light illuminated in the respective fire shut-off handle and the fire warning bell rang. Pulling out the handle, among other actions, closed the fuel shut-off valve to that engine. As it also electrically armed the fire extinguisher discharge switch, the fire extinguisher bottle could only be discharged after the handle had been pulled. Examination of No 2 engine's two fire extinguisher bottles showed that neither was fired electrically from the flightdeck.

The reason for this lay with the crew's failure properly to complete the fire drill actions prescribed in BOAC's emergency checklist for the Boeing 707. The emergence of a sudden engine fire of growing intensity meant that engine fire drill quickly supplanted engine failure drill, but apart from silencing the fire warning bell, the two drills differed in only one crucial respect — the pulling of the fire shut-off handle. As it happened, this difference was inadvertently obscured by Flight Engineer Hicks when he initially reached for, but did not actually pull, the shut-off handle when carrying out the failure drill. This evidently gave both him and First Officer Kirkland the impression that the fire shut-off handle had already been pulled when Capt Taylor called for the fire drill only a short time later. AIB investigators believed that Capt Taylor was so preoccupied with physically handling G-ARWE that he had no spare capacity to monitor his crew's performance. Understandably in the circumstances, he relied on their report that the fire drill had been completed.

Check Captain Moss did not monitor the fire drill either because he was watching the progress of the engine fire and helping Taylor position the aircraft for landing. Furthermore, with the physical loss of No 2 engine, the red warning light in the shut-off handle would have gone out, which may have convinced the flight crew that everything that needed to be done had been done.

And then Sod's Law came into play. It was evident to the accident investigators that the fire warning bell did not ring when the engine fire broke out because Flight Engineer Hicks, hearing the undercarriage warning horn a moment before, had misidentified the action to be taken and pressed the fire bell cancel button at the instant the bell would have begun to ring. Had it actually rung, and his first action had been an intentional cancellation of the warning bell, it was highly likely that there would have been no confusion over completing the other memory items from the fire drill checklist.

Overall, this accident resulted from an untimely combination of unusual circumstances and unfortunate coincidences. In among these was the presence on board Flight 712 of Check Captain Moss. BOAC's policy on such

Opposite: *Boeing 707 crew positions on the flightdeck of G-ARWE.*

routine checks was that the 'check captain... is expected to be as unobtrusive as possible'. It would have been impossible to remain totally 'unobtrusive' in the circumstances that befell Flight 712, but it was probably Moss's action in silencing the undercarriage warning horn when the flight engineer began the engine failure drill that engendered the initial confusion in Hicks' mind, leading him inadvertently to silence the fire warning bell. This in turn led to the most tragic omission of all — the failure to pull No 2 fire shut-off handle. 'Too many cooks,' be it in the kitchen or on the flightdeck, can disrupt the most well-tried routines and generate an atmosphere of urgency which begets confusion.

All of which showed that there are degrees of remaining in control of an aeroplane. Despite an excellent approach and landing by Capt Taylor in the circumstances, the growing ferocity of the fire countered most of the advantage gained by skilful handling in the air. The end result of a marked lack of crew co-operation was the needless loss of five lives and a perfectly good aeroplane.

The great problem with fire in aeroplanes is that it kills in a unique fashion. Despite the evocative pictures of the airship *Hindenburg* crashing in 1937, people hardly ever die from being burnt to death at altitude. Where most fires kill is on the ground. Modern airliners are pretty rugged beasts, and it can be seen from this book that aircraft can be pretty thoroughly wrecked and still leave survivors. But if a wreck catches fire on the ground, and hundreds of duty-free-loaded tourists stumble around and inhibit an efficient evacuation, they can all die from suffocation before they get out.

On 19 August 1980, a Saudi Arabian Airlines (Saudia) Lockheed L-1011-200 TriStar (HZ-AHK) had arrived at Riyadh International airport from Karachi. It was late in the evening when Flight 163 started off from the Saudi capital on the second leg of the scheduled service to Jeddah. Only seven minutes after take-off, HZ-AHK's CVR showed that the flight crew was alerted by both visual and aural warnings to the presence of smoke in the aft cargo compartment, designated C-3. The initial alert occurred as the TriStar was climbing through 15,000ft. The crew spent more than four minutes trying to confirm the warning and looking up the smoke warning procedure. The captain then decided to return to Riyadh, which was just as well because confirmation of the fire came as the airliner was on its way back to the theoretical safety of terra firma.

From then on, the situation only got worse. The commander failed to use his first and second officers, especially the former to whom he should have delegated the task of flying the TriStar while he concentrated on the wider implications of the emergency at hand. Perhaps the commander never accepted that there was a real emergency, an impression reinforced by the flight engineer who failed to provide an accurate picture of what was happening and kept saying, 'No problem,' when in fact a serious one existed. The commander was not the first in history to try and do everything himself while the first officer, who had only limited experience on the aircraft type, did not assist the commander in monitoring the safety of the flight. The second officer did not help matters by confusing both his instruments and procedures.

None of the flightdeck trio was affected by the smoke that was filling the passenger cabin until after landing. In contrast, the cabin staff behaved commendably, both in battling the blaze and trying to calm their panic-stricken passengers. It is unlikely that the cabin staff could have done much in the circumstances to aid the eventual evacuation of the aircraft, assuming that the captain had made any preparations for such an eventuality.

Notwithstanding a jammed thrust lever, which necessitated shutting down the TriStar's No 2 (centre) engine, HZ-AHK landed safely. But then the captain made another critical error. Rather than use maximum available braking power to stop the TriStar expeditiously in order to get everyone out, he continued to taxi off the runway. Perhaps he did not want to close down the airport by blocking its in-use runway, but he did not bring the airliner to a stop on the taxiway until 2min 40sec after touchdown. Moreover, by keeping Nos 1 and 3 engines running for another 3min 15sec, not only did he prevent the airport emergency services from taking immediate action but he also thwarted any attempt by his cabin crew to initiate their own emergency evacuation.

The last message from Flight 163 came at 21.40hrs: 'We are trying to evacuate now'. It was all too late. There was no evidence that anyone inside the stricken airliner tried to open the cabin doors. Maybe passengers were blocking the doors, which have to move a few inches inward to open, but it was more likely that both flight and cabin crew were incapacitated by a flash fire which burnt up whatever oxygen remained inside the TriStar. The situation was not helped by the fact that the environmental control system packs, which brought fresh air on board, had been switched off on landing in accordance with normal procedures. No one up front seemed to have appreciated that the situation facing Flight 163 was anything but normal.

The airport fire and rescue services were also woefully inadequate, lacking both correct equipment and proper training to cope in these circumstances. It took them 23min

Below:
The remains of Saudi Flight 163 after the TriStar had been gutted by fire at Riyadh on 19 August 1980.

after engine shutdown to get into the fuselage, by which time all 301 people on board Flight 163 (287 passengers and 14 crew members) had long since perished. The fire ultimately consumed almost the entire upper fuselage structure wherein sat all the passengers and crew, but they died not so much from burns as from the inhalation of such toxic nasties as carbon monoxide, nitrous oxide, hydrogen cyanide, formic acid and ammonia, not to mention oxygen starvation.

The origin of the fire could not be determined as the source was obliterated in the blaze. It could have come from something as simple as the accidental ignition of matches left in a suitcase, but there was no evidence of any incendiary device. That said, the subsequent accident investigation was in no doubt that the fire had started in compartment C-3. The throttle controls cables, suspended from fairlead rollers, were routed through the space between the cabin floor and the top of the cabin hold. These rollers softened, melted and adhered to the control cables when heated, which accounted for the sticking of No 2 thrust lever.

Smoke and flames would have travelled the same path to the sidewall of the aircraft and up into the passenger cabin. The burning of furnishings and other cabin materials only added to the noxious cocktail. Lockheed made several modifications to the TriStar fleet following this tragedy but these were largely palliatives. In the final analysis, it was not HZ-AHK which failed at Riyadh. Notwithstanding the fire in the aft cargo compartment, the TriStar was robust enough to keep flying until completion of a successful emergency landing. Yet in dithering about after landing, the flight crew allowed HZ-AHK's furnishings to burn to the point where they not only consumed oxygen but also gave off poisonous gases and floating solid debris which blocked mouths, noses and, just as crucially, eyes. The physical and psychological effects of dense smoke in a tightly packed cabin interior must have increased fear and blind panic. The fact that 301 people died in consequence when supposedly safely on the ground, making the loss of Flight 163 one of the worst disasters in the history of commercial aviation, showed that the modern threat from fire is more than just excessive heat.

11 Cockpit Indiscipline

In October 1932, the Soviet leader Joseph Stalin decided that a special aeroplane should be built for propaganda purposes. Designed by a team headed by Andrei Tupolev, the result was the largest aircraft in the world for its time: the huge, eight-engined, all-metal Ant-20, with a 206ft wing span, manned by a crew of eight and which could carry 72 passengers. But passenger flying was not the main aim; for showing the party flag, the Ant-20 had equipment unmatched by any aircraft before or since. The fuselage was large enough to accommodate a cinema, a projection cabin for films and transparencies including the facility for in-flight projection on to clouds or artificial mists, a complete photographic and film laboratory, an offset printing office, an internal telephone system for 12, electric fluorescent writing on the underside of the wing, and a gigantic external loudspeaker through which music and inspirational messages could be blasted out over an area of 12sq km. The Ant-20, which first flew on 17 June 1934, was named *Maxim Gorki* after the popular writer and handed over to the *Maxim Gorki* propaganda squadron on 18 August. During subsequent advertising flights, newspapers were printed and then thrown out as the massive aircraft passed by. Not surprisingly, *Maxim Gorki* always aroused a great deal of attention, though in the end for all the wrong reasons.

On 18 May 1935, the flying mammoth crashed at the Tushino airfield near Moscow after an escorting Polikarpov I-5 fighter hit the *Maxim Gorki*'s wing during promotional filming. The little biplane fighter was being flown alongside for visual contrast when its pilot, Nikolai Blagin, ignored orders from the chief pilot of the Gorki Squadron, I. V. Mikheev, to keep his distance, saying over the radio, 'I'll show you how good a flyer I am'. Blagin tried to loop around the far bigger machine, but he crashed into it instead, sliding along the fuselage and into the rudder. Both aircraft hit the ground in spectacular fashion, killing Blagin and all 44 people on board the *Maxim Gorki* including Mikheev and the captain and pilot, I. C. Shurov.

It was as well that Blagin died in the accident he caused, as it would have been no fun to face the wrath of General-Secretary Stalin after the subsequent board of inquiry. What was intended as a showcase for the Soviet political system, its technology and its aviation industry had turned into a public relations disaster in a matter of seconds. And all because of human indiscipline in the air.

Notwithstanding all the in-depth training, regular check-rides and psychological profiling, aircrew can still behave in the same foolhardy or irrational manner as other human beings. On 16 July 1957, a Lockheed Super Constellation (PH-LKT) belonging to KLM Royal Dutch Airlines took-off from Mokmer airport on Biak Island in Netherlands New Guinea. PH-LKT was on a scheduled service to Amsterdam, but first the pilot asked for clearance to make a low pass over Mokmer. No reason was given to ATC but the captain announced over the PA system that he intended to give his passengers a final glimpse of the island before setting off on the first leg to Manila in the Philippines.

On being granted air traffic clearance, the Super Constellation began its approach from the east, gradually descending over the sea. It crashed approximately half a mile from the shore, burst into flames, broke apart and sank. Of the 68 people on board PH-LKT, only 10 survived. The time of the crash was 03.36hrs, and it was more than likely that the pilot misjudged his height in the dark, moonlight conditions. What glimpses of the island the captain hoped to show his passengers at that time of night we will never know, but he was not the first to forget that there is a wealth of difference between airline operations and display flying.

There are many understandable reasons why flight crews bend the rules. For example, in summer 1995 a Britannia Airways Boeing 767 left Manchester airport bound for Ibiza to pick up a group of tourists. There was only the crew on board but instead of flying directly to sunny Spain, Capt Hugh Carmichael made an *en route* diversion to overfly the Cheshire town of Congleton. The 2,000ft flypast was intended as a surprise for First Officer Michael Stanley's baby daughter who was attending a birthday party. But other residents of Congleton were not impressed, and after an inquiry into the use of a 'non-standard manoeuvre,' Capt Carmichael resigned from the airline on 16 November 1995, severe disciplinary sanctions were imposed on his first officer, and Britannia apologised to the citizens of Congleton. In exposing their aircraft, those on board and those below to unnecessary risk, both KLM and Britannia captains made a human misjudgement for which they both paid a high price.

The reasons behind some human failures can be tragic in themselves. On 9 February 1982, a Japan Air Lines DC-8 hit the sea some 900ft short of Tokyo's Haneda airport, killing 24 of the 174 on board. Investigators subsequently found that Capt Seiji Katagiri, who had known mental problems, may have had a breakdown and put an engine into reverse while the co-pilot and flight engineer battled to restrain him. It is easy to say that Capt Katagiri should have been grounded well before 9 February, but often there is no clear dividing line between counselling and destroying a career.

Any airline training and checking system worth its salt ought to be able to guarantee that its flight crews know what they should be doing. On 8 November 1961, a Lockheed 049E Constellation (N2737A) belonging to Imperial Airlines was transporting 74 US Army recruits from various pick-up airports to Fort Jackson near Columbia, South Carolina. Although this was a military contract, the flight crew of five (including a student flight engineer) were all civilians. As the Constellation took off from its final pick-up point at Baltimore, a momentary fluctuation was noted in No 3 engine fuel pressure. This was probably caused by the failure of its fuel booster pump — each Wright Double Cyclone engine had direct fuel injection — but when the student engineer opened Nos 3 and 4 cross-feed valves, the higher pressure from No 4 booster pump held closed a check-valve between the cross-feed manifold and No 3 fuel tank. Consequently, No 4 tank was left to supply both starboard engines on its own; it was soon exhausted, causing both 3 and 4 engines to fail.

The regular flight engineer then took over but immediately he made a procedural error by leaving No 4 booster pump on; this allowed air to be drawn into the fuel lines supplying both starboard engines, thereby preventing either from being restarted. The crew then elected to divert to Byrd Field just outside

Richmond, Virginia. During the approach to Runway 33, the commander (who was acting as co-pilot) inexplicably decided that they should switch to Runway 03. The landing then had to be abandoned because the undercarriage would not lower. The gear was normally activated by pressure provided by hydraulic pumps driven by the starboard engines. As they had failed, the crew should have activated the crossover valve that would have enabled the hydraulic pumps driven by Nos 1 and 2 engines to assume the task. The flight crew omitted to make the appropriate selection.

As the Constellation circled Byrd Field in an anticlockwise direction with a big thinks bubble overhead, the commander decided to revert to a landing on Runway 33. But during a poorly executed overshoot, the flying pilot allowed the bank angle to steepen to such an extent that the airliner lost speed and then altitude. In an attempt to counter a high sink rate, the crew applied full power to the port engines but under the strain of overboosting, No 1 failed. Unable to maintain height on only one 2,200hp engine, the Constellation finally crashed approximately one mile from the upwind runway threshold. A sharp pull-up was made just before the airliner flew into woods, but although the crash was survivable, 77 people died from carbon monoxide poisoning in the subsequent fire. Only the commander and the regular flight engineer survived.

The crash occurred in darkness around 21.30hrs but the weather was good. The subsequent investigation attributed the accident to a lack of command co-ordination and decision (as illustrated by the captain's decision to change runways), faulty judgement and insufficient knowledge of aircraft systems; the combination created an emergency situation that the crew could not handle. In addition to lack of aircrew competence, other irregularities were discovered including the finding of rust in the Constellation's fuel system and the tanker that served it. Although this contamination did not cause the multiple engine failure, it was symptomatic of the company's lack of compliance with civil aviation regulations. Consequently, the FAA revoked Imperial Airlines' operating certificate six weeks after the crash. There can be no place in the air for people who don't know what they are about.

One flight crew particularly lacking in situational awareness operated an Eastern Airlines DC-9 (N8984E) on 11 September 1974. The accident happened as the crew of Flight 212 was conducting a VOR/DME instrument approach to Runway 36 at Douglas Municipal airport near Charlotte, North Carolina. With undercarriage down and flaps set at 50°, the DC-9 hit trees at around 07.35hrs before crashing into a field some three miles short of the runway. Upon impact, the airliner broke up and burst into flames, killing 72 of the 82 people on board including the captain.

Although the loss of Flight 212 was not the worst accident of the year, it was among the most notorious. There was shallow, patchy fog in the area at the time of the accident, but that did not seem to bother the two pilots. The CVR tape revealed that up to about 2.5min before impact, they were engaged in a conversation that had nothing to do with the operation of the aircraft. They covered a variety of topics, ranging from politics to used cars, with both pilots expressing strong views and mild aggravation throughout. In the process they relaxed their instrument scan, relying more on visual cues to carry out the approach. This was fine until N8984E entered the fog; thereafter, there was insufficient time to revert to instrument procedures.

Flight 212 passed over the final approach fix some 450ft low and well above the recommended speed. The DC-9's terrain warning system sounded at 1,000ft above the ground but the crew did not heed it. The first pilot was doing the flying, but his captain omitted to make the mandatory check-calls when reaching 500ft above airport elevation or 100ft above minimum descent altitude. And the captain did not help himself by mis-setting the pressure on his altimeter such that it over-read by nearly 800ft. All in all, the airliner was lost because of a series of errors caused mainly by lax cockpit discipline at a most crucial stage in flight.

Airline management has a responsibility to ensure that all in their employ who are closely associated with aircraft operations are fit to do their job. There must never be a repeat of the loss of DC-4 (CF-MCF) belonging to the Canadian airline Maritime Central Airways, which crashed near Issoudun, Quebec, on 11 August 1957. The airliner was bringing back World War 2 veterans and their families from a trip to England, and after a long transatlantic haul, the airliner entered a thunderstorm while flying at 6,000ft. Suddenly, it diverged from controlled flight and plunged to earth, bursting into flames on impact and killing all 79 people on board. The cause of the crash was never positively determined, but if the pilots had been guilty of any misjudgements, it was probably because they had been on duty for nearly 20 hours. The resulting fatigue must have affected their ability to cope with any adverse situation, and thereafter all airlines introduced on-duty time limitations for their flight crews.

But notwithstanding all the legislative and human resource improvements over the last 40 years, cockpit indiscipline remains an abiding flight safety concern even in the most strictly regulated organisations. At about 14.00hrs on 24 June 1994, a B-52 bomber took off from Fairchild Air Force Base in Washington State to practise an air show display. Barely 15min later, while attempting to circle the tower in a steep turn, the great machine crashed, narrowly missing a nuclear weapons storage bunker and a crowded airmen's school.

The pilot, Lt-Col Arthur Holland, was about to retire having been in the USAF for 24 years. But experience had not mellowed him for he had a reputation as a 'hot stick'. He once climbed the great 185ft-wing span 'Buff' so steeply that fuel flowed out of the vent holes on top of the wings. His hard flying during one air show popped 500 rivets during a prohibited manoeuvre, and he put another B-52 into a 'death spiral' over one of his daughter's high school softball games. A co-pilot complained that he had to wrestle control from Holland on one occasion, and three months before his fatal flight he cleared a ridgeline by only a whisker. In the end, hardly anyone would fly with Holland; two of the three officers who died on 24 June were there because their subordinates feared to fly with him. Yet in spite of all this, Lt-Col Holland was put in charge of evaluating all B-52 pilots on the base.

The Holland saga was one of some 30 cases that former USAF chief civilian safety officer Alan Diehl made public in May 1995 in an effort to expose deteriorating safety standards in the US Air Force. On another occasion, two Navy fighter pilots and a navigator removed their clothes, helmets and oxygen masks in an attempt to 'moon' the crew in another aircraft. The three men died when their mismanaged aircraft crashed. And in 1981, a KC-135 (the military version of the Boeing 707) spiralled down from

Opposite: *Lt-Col Holland's B-52H comes to grief at Fairchild AFB on 24 June 1994. The 'Buff' was too big to mess about with.*

over 25,000ft into a barley field after two wives — who were accompanying their husbands on a 'spousal orientation flight' — were allowed to sit in both pilots' seats and then fiddle around to such an extent that the KC-135 went out of control.

Perhaps such flagrant breaches of cockpit discipline are indicative of the modern age, where aviation and aircraft systems have become so relatively commonplace, reliable and sanitised that pilots feel the need to kick over the traces every so often to rediscover that frisson of excitement which was once the lifeblood of magnificent men in their flying machines. Whatever the stimulus, there is no doubting that indiscipline crosses the aviation spectrum. In 1994, the world total of fatalities from airline accidents rose by 25% to 1,385 compared with 1993. These losses flowed from 47 accidents, and aircrew error was a factor in 31 (or 67%) of them. It would be wrong to imply that they were all in the same category as Lt-Col Holland's transgression. For example, on 6 June 1994 a China Northwest Airlines' Tupolev Tu-154M (B-2610) medium-range airliner went out of control most dramatically when its autopilot was engaged eight minutes after take-off. It flew into the ground near Xian killing 160 people; their death was subsequently found to have been caused by faulty cross-wiring of the autopilot channels during pre-flight repairs. But if this accident was caused by laxity on the ground, there are still too many clowns flying about for comfort.

On 22 March 1994, Airbus A310-300 (F-OGQS) in the colours of the Russian international carrier Aeroflot was cruising at 33,000ft on autopilot *en route* from Moscow to Hong Kong at night. It was crossing Siberia some four hours into the flight when the airliner's trace suddenly disappeared from ground radar screens without so much as a distress call. FDR data indicated a dramatic spiral descent with bank angles exceeding 90°, two rapid pull-ups and stalls. Just before impact, the airliner nearly regained a stable flightpath via a 4.8g pull-up but not quite; the Airbus crashed at Novokuznetsk, killing all 75 people (63 passengers and 12 crew) on board.

Subsequent investigation of this bizarre accident indicated that Capt Yaroslav Kudrinski's son and daughter were invited on to the flightdeck. Capt Kudrinski allowed his 15-year-old son to sit in the left-hand seat whereupon the commander began instructing the teenager in flying techniques. The boy apparently turned the control yoke to the right, placing the airliner in a slight right bank. But then the bank angle rapidly increased in excess of 90° and the nose dropped sharply. When this happened, someone pulled the control column back in an attempt to regain level flight, but the nose was pulled too high and the A310 stalled.

Rather amazingly, it appears that instead of the pilot sitting in the right-hand seat taking over control immediately at the first hint of trouble, CVR recordings indicated that the teenager was being coached by Capt Kudrinski in recovery actions from incipient unusual positions. Not surprisingly the lad made a right hash of it, with the airliner being stalled several times. The Airbus's nose was pulled as high as 50° above the horizon during at least one of the attempted recoveries. What the pilot in the right-hand seat was doing to let all this happen beggars belief.

In the end, the FDR showed that he must have initiated the 4.8g pull-up which almost did the trick; the airliner had a minimal descent rate when it hit the ground in a slightly left-wing down attitude. It said a great deal for the Airbus's structure that it remained intact during the high g manoeuvre, but it said very little for the pilots' sense of duty and situational awareness that they treated their aircraft and passengers in such an irresponsible manner.

On the face of it, such a disaster should have been ruinous to Aeroflot's marketing strategy. So it may have been, but ironically, the intense internal and external scrutiny which was then focused on the airline may well have made Aeroflot into one of the safest airlines on which to travel, at least in the short term. But that outcome should never be allowed to mask the fact that a consistent safety record is the best selling point for any airline. In any first rate aeronautical organisation, there can be no place for second rate attitudes.

Left:
The lengths to which accident investigators sometimes have to go to get at the cause of an accident. In October 1965, a BEA Vanguard crashed at Heathrow. Its reassembled remains are being inspected here by the accident inquiry team seven months later at the Royal Aircraft Establishment, Farnborough.

12 Happy Landings

When an airliner starts to descend, passengers' stomachs tend to sink too. On the face of it, they have just cause. The 25 December 1946 was nicknamed 'Black Christmas' in China as three airliners crashed in succession while trying to land at Shanghai in bad weather. Over 40% of flying accidents still happen during the relatively short time from final approach to touchdown. Yet the bad old days of lumbering airliners descending through clouds with little more to guide them than a plumb line and a prayer have long since gone. In terms of deaths per hundreds of thousands of flying hours, even approach and landing accidents are now rare. For that, travellers must partly thank the instrument landing system (ILS), which was adopted by the International Civil Aviation Organisation (ICAO) as a worldwide standard in 1949.

ILS allows aircraft to land even when weather conditions prevent the flying pilot from seeing beyond 250ft ahead. ILS signals come from two radio transmitters on the ground, each working a different frequency. The 'localiser' transmitter at the far end of the runway sends a beam centred along the runway; this shows whether the aircraft is off to the left or to the right. The 'glidepath' transmitter, close to the point where the aircraft will touch down, points a beam upwards angled at 3^{0}; this tells the crew whether the aircraft's descent rate is just right, too steep or too shallow. In essence, any pilot who keeps the ILS glidepath and centreline bars crossed in the centre of his instrument will miss all obstacles on the approach and emerge from bad weather over the white piano keys painted on the runway threshold. Indeed, ILS signals can be fed into the flight computer so that an airliner can land itself.

Above:
A Handley Page Hermes runs off the end of Southend airport runway after returning from Majorca on 9 October 1960. At least it met the requirements of the old adage that 'a good landing is one you can walk away from'.

But technology can still founder in the face of human frailty. N1996 was a Boeing 727 delivered to American Airlines in the summer of 1965 as part of American's transition from piston-engined airliners to jets. It had less than 1,000 flying hours on the clock when it was assigned to American Airlines Flight 383, scheduled to depart New York LaGuardia at 17.00hrs on 8 November 1965 bound for Cincinnati. Capt Bill O'Neill, a 39-year-old pilot with 14,000 flying hours under his belt, was in the left-hand seat. He had been with American Airlines for 14 years, flying piston-engined airliners hither and yon. O'Neill had completed a conversion course on to American's new Boeing 727 trijets three weeks earlier. By the evening of 8 November, he was well into the 25 hours of line flying necessary to confirm captaincy on type under the supervision of Check Captain David Teelin, 46, who would be carrying out the duties of first officer from the right-hand seat on the 90min leg to Ohio.

The 727 eventually left LaGuardia 20min late because of N1996's delayed arrival at New York from an earlier flight. The terminal forecast for Cincinnati read: 'Ceiling 1,200ft broken, 3,500ft overcast, visibility four miles, light rain, fog. Variable to 1,000ft overcast, visibility two miles, thunderstorms, moderate rain.' By transitting at FL350 and dog-legging south over Charleston, West Virginia, the 727 kept clear of weather trouble *en route*.

The crew checked in with Greater Cincinnati airport at 18.45hrs and 10min later, when the 727 was 24nm from the airport, it was transferred to Cincinnati Approach Control. Darkness was falling but Flight 383 was in clear weather to the east of the airport. Consequently, the crew opted for a visual approach: 'Out of 5,000 for 4,000 (feet) — how about a control VFR? We have the airport in sight.' Although there was cloud and some lightning around the airport to the northwest of them, a visual approach would save time and claw back some of the delay incurred by the late departure from LaGuardia and the *en route* diversion.

The approach controller cleared Flight 383 for 'a visual approach to Runway 18, precip lying just to the west boundary of the airport and it's southbound'. This instruction was acknowledged and the pilots cleared to descend to 2,000ft at their discretion. Three minutes later, the controller put N1996's position at 6nm southeast and told them to call Cincinnati Tower:

Above: *'Please be patient while we wait for the steps to arrive.' A Vickers Varsity gives the owners of this house in Gloucester a fright on 27 March 1963.*

Opposite: *An American Airlines 727-23, identical to that which crashed while attempting to land at Greater Cincinnati.*

Flight 383: 'Cincinnati Tower... we're 6 miles southeast and... control VFR.'
Tower: 'Runway 18, wind 230°, 5kt, altimeter (QNH) 30.'
Flight 383: 'Roger, Runway 18.'
Tower: 'Have you in sight — cleared to land.'
Flight 383: 'We're cleared to land, roger. How far west is that precip line now?'
Tower: 'Looks like it's just about over field at this time, sir. We're not getting anything on the field however... if we have windshift, I'll keep you advised as you turn on to final.'
Flight 383: 'Thank you — we'd appreciate that.'

Ten seconds later (at 19.00:13hrs), the 727 reached the end of the downwind leg and turned towards the airport.

Tower: 'We're beginning to pick up a little rain right now.'
Flight 383: 'OK.'

A minute later.

Tower: 'Have you still got the runway OK?'
Flight 383: 'Ah... just barely... we'll pick up the ILS here.'
Tower: 'Approach lights, flashers and runway lights are all on high intensity.'
Flight 383: 'OK.'

Nothing more was heard from Flight 383. It was then crossing the centreline of Runway 18 barely two miles from touchdown, but notwithstanding proximity to the runway threshold, whatever sighting Capts O'Neill and Teelin once had of the reassuring concrete had been lost in mist and low cloud. Vainly they sought to reorientate themselves by picking up

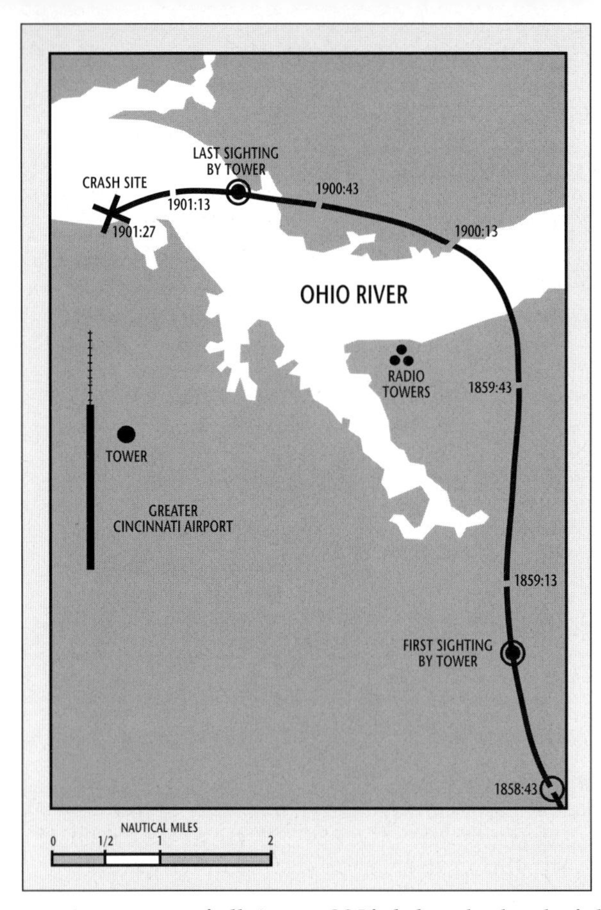

the ILS, but to no avail. At 19.01:27hrs, five seconds after the last radio transmission, the perfectly serviceable N1996 flew into the wooded west bank of the Ohio River valley. After cutting a 300m swathe through scrub and foliage, the 727 collided violently with a clump of trees and burst into flames. Fifty-eight people including five crew members died in the process. Only one stewardess and three passengers, one of whom was an off-duty American Airlines pilot, survived.

Greater Cincinnati airport is 890ft above sea level. Examination of the wreckage showed that when disaster struck, N1996 was about 400m to the left of the Runway 18 extended centreline, it was on a heading of 235° in a level attitude, but, most important of all, it was 225ft *below* the level of the airport.

All airfield lights had been at their highest intensity during the 727's approach, and the ILS was fully serviceable. Given that the undercarriage was still retracted at the moment of impact, everything pointed to the accident having resulted from the failure of the flight crew to monitor properly the aircraft's height during the final approach. But why did two very experienced pilots appear to have let their 727 go through a final descent lasting nearly 120sec without sufficient reference to their altimeters?

American Airlines practice was to set both pilots' altimeters to airport QFE, while leaving the centre altimeter set to QNH. N1996's three Kollsman drum pointer type altimeters were examined by the manufacturers; nothing untoward was found and the settings on both pilots' instruments approximated to the airport QFE setting passed by Air Traffic Control. Evidence from witnesses and the airliner's FDR showed a continuous descent from 7,000 to 2,000ft on a northwesterly heading of 305°, during which the airspeed progressively reduced from 350 to 250kt. At 2,000ft (1,100ft above airport elevation), the 727 levelled off to fly north on the downwind leg while the airspeed bled off to 190kt. It then entered a gentle left turn on to base leg. At this stage it began descending again at a steady 800ft/min, with the airspeed decreasing to 160kt. Half a minute before impact, the 727 recommenced the left turn on to finals. Within 10 seconds, the rate of descent had increased to just over 2,000ft/min, taking the 727 below the level of the airport into the Ohio River valley. During the final 10 seconds, the descent rate eased off to 625ft/min and the speed to 147kt. Whatever else the pilots were doing, they were not looking at their instruments.

Several interrelated factors were thought to have brought this about. First, the pilots were rushed. Capt O'Neill was doing the flying while Capt Teelin assumed the role of co-pilot. Post-crash investigators gained the impression that the entire flight to Cincinnati was conducted with the aim of getting the 727 down on the ground as quickly as possible. This may have been to catch up lost time, and also to get in ahead of the deteriorating weather, but the upshot was that the airliner's average groundspeed within the Cincinnati terminal area was in excess of 325kt when it should have been no more than 250kt.

Consequently, flap extensions were bunched up on base leg and even then, the crew only managed to get below flap limiting speeds by flying the entire base leg descent at or near engine idle. Yet according to American Airlines' procedures, when the 727 turned on to final approach it should have been in the full landing configuration with undercarriage lowered, 40° of flap extended, and airspeed and rate of descent stabilised so that only small adjustments to the glidepath, approach speed and trim would be necessary. But because higher than normal speeds had been flown up to then, only 25° of flap had been extended, the undercarriage was still up and a number of other landing checklist items remained to be completed. With less than two miles to go to the threshold, the crew would have been working like one-armed paperhangers.

If only they had extended the downwind leg, everything could have been accomplished in a much more controlled fashion. Thereafter, with aircraft slowed and the proper degree of flap extension, Capt O'Neill could have used higher thrust settings, allowing him to control the approach and complete the landing checklist in an unhurried fashion.

Having got themselves behind the mental drag curve, Capts O'Neill and Teelin were not best placed to counter a couple of 'gotchas' that crept up unannounced. The first was the weather. As the crew turned on to base leg, they left VFR conditions behind and ran into light rain and low scudding cloud which rapidly reduced visibility. At this point, the pilots might have felt that they had to descend to maintain visual conditions.

But the rain became heavier around base leg and visibility dropped to two miles or less. In such circumstances, only the sequenced 'flashers' of the airport approach lighting system would have remained visible. The pilots may have become preoccupied in looking out to the left of the airliner, striving to keep these in sight. However, at this stage the Ohio River valley lay just to the left of the flightpath. It was 400ft lower than the airport, and probably the only lights in the pilots' field of view were those of houses on the river bank. In poor visibility, these lights could have given the pilots an illusion of adequate altitude above the airport.

The drum pointer altimeters were fine in general use, but investigators found that, under conditions of infrequent or distracted altitude monitoring, a misinterpretation of negative for positive height readings could occur. That should not have worried two experienced captains, but maybe they got out of alignment too. It was possible that

Opposite: *The final flightpath flown by American Flight 383 into Cincinnati on 8 November 1965. The 727 came down near Constance, Kentucky.*

Capt Teelin, confident of his 'pupil's' abilities, assumed that Capt O'Neill was monitoring the instruments while he concentrated on keeping the airport control lights in sight. O'Neill on the other hand, who was in control, may have taken comfort from the fact that he had an experienced check captain in the right-hand seat who would be monitoring the instruments, leaving O'Neill to concentrate on looking for the approach lights.

As the pilot not making the landing, Capt Teelin had a high workload of extending the flaps, running through the landing checklist with the flight engineer, and making radio transmissions. But as he was also concentrating on locking on to the airfield approach lights out to the left of the airliner, it would not have been surprising if he had little opportunity to swing his gaze back inside the cockpit to focus on his own instrument panel on the right-hand side of the cockpit. It is more than likely that he used O'Neill's altimeter on the left-hand side of the cockpit for periodic reference; unfortunately, it is well known that the probability of error is increased when an instrument is read from a side angle.

But Capt Teelin was missing more than that. American Airlines required their non-flying pilots to call airspeed, altitude and rate of descent whenever an aircraft descended to 500ft on final approach. Rate of descent was to be called again if at any time it exceeded 700ft/min thereafter. N1996's FDR showed that the 727 descended through 500ft on base leg 42sec before impact, and that the rate of descent remained in excess of 700ft/min throughout the remainder of the approach. Teelin must have been so distracted in looking for the approach lights-cum-reorientating with the ILS that his instrument scan broke down completely.

In ideal circumstances, Capt O'Neill on the left-hand side of the cockpit should have concentrated on looking for the approach lights out to the left in a left-hand turn, leaving the 'poling' to Teelin. But by the time it all went pear shaped, it was too late to swap control in any safe fashion, assuming that anyone on the flightdeck had the spare mental capacity to come up with such a course of action.

The last point at which the accident could have been averted was 13sec before impact when the 727 was descending below the level of the airport. Capt O'Neill could have cranked up the power and carried out a missed approach procedure into a full ILS but he did not, and final visual contact with the airfield lights was lost.

It should serve as a salutary lesson to all aircrew that on 8 November 1965, two highly experienced and respected captains spent nearly two minutes descending from only 1,200ft above Cincinnati airport at night in adverse weather without adequately monitoring their altitude. And all because, in their haste to complete their flight, they let themselves get into a situation whereby, when the weather suddenly worsened, they became too distracted to realise

Opposite above: *The Convair 880 tail assembly at the crash site near Covington, Kentucky. Often the tail section is the only major structure that remains after an airliner crash, and if you want to increase your chances of survival should that one-in-a-million crash occur, book a rear seat. The trouble is, that may put you in the smoking section, and you then have to balance one risk to your health against another.*

Opposite: *A sister aircraft to this TWA Convair 880 came to grief while trying to land at Greater Cincinnati on 20 November 1967.*

TWA
N821TW
880

TWA
N815TW
TWA

how close they were to disaster. Little things like an ergonomically-challenged altimeter can push pilots over the edge when they reach the limits of their thinking time.

Airline management also have a responsibility to ensure that crews constantly exercise a conservative and prudent approach to their daily work. In other words, while flight crews have to obey the rules, airline management must generate a corporate culture that is not so hung up on the bottom line that, by implication, it positively encourages the cutting of corners.

In concluding its accident investigation, the board re-emphasised that the responsibility and authority given to airline captains required the continual exercise of sound judgement and strict adherence to prescribed operational procedures. Any deviation from such standards could only compromise aircraft safety. The need for such advice was borne out almost exactly two years later when a four-engined Convair 880, making a night approach in reduced visibility into Cincinnati, struck trees and came to grief just short of the Runway 18 threshold in remarkably similar circumstances to those that befell Flight 383.

Weather plays a big part in aircraft landing accidents but really this is only the trigger. On 10 April 1973, a Vickers Vanguard four-engined turboprop (G-AXOP) belonging to Invicta International Airlines was operating on charter flight IM435 from Luton via Bristol to Basel-Mulhouse airport on the French side of the Franco-Swiss border. 139 passengers were taken on board at Bristol, whereupon the Vanguard took off for Basel at 07.19hrs with Capt Anthony Dorman in the left-hand seat and Capt Ivor Terry (the non-flying pilot in charge of the radios) in the right.

At 08.49hrs, IM435 reported inbound for BN, the Basel main approach beacon. The weather was passed as 'Wind 360/9kt, RVR (runway visual range) 700m, snow, cloudbase 8/8 120m, QFE 967.5mb, QNH 998.5mb, temperature 0°, runway in use 34'. Flight IM435 acknowledged and called Basel Tower which instructed the crew to maintain FL70 and report passing BN. In response to a query from the crew, Runway 16 was allocated for the approach.

At 08.55:48hrs the pilot reported passing the BN beacon, which also served as the ILS outer marker. He was cleared to descend to 2,500ft, the initial altitude for an instrument approach, with an instruction to report when settled into the first procedure turn. This was confirmed at 08.57:42, whereupon permission was given for the final approach. At 09.00:13hrs, Capt Terry reported passing BN; tower passed clearance to land on Runway 16 and a surface wind of 320/8kt.

At 09.05:12hrs Capt Dorman reported that G-AXOP was overshooting for another approach, which was cleared. At 09.07:27 the crew said that they had passed BN outbound, which implied that they were progressing north-northwesterly. But 43sec later, Basel Tower received a telephone call from a meteorologist and former aircraft commander who stated that, barely two minutes before, a four-engined turboprop aircraft with red vertical tail surfaces had flown over the Binningen Observatory (some 8km southeast of the airport) at around 150ft above the ground and heading south. He urged that the crew be told to climb.

At 09.11:10hrs, Zurich Area Control asked Basel whether they had an aircraft which was flying outbound towards Hochwald in Switzerland as they had an unidentified blip on their radar 3-5nm southwest of Basel. The Basel controller denied this, but on checking his own scope he observed a clean echo on the extended runway centreline approximately 6nm south of the airport heading south. Fifteen seconds later, IM435 reported that it was over BN to the north and was once again given permission to land. At 09.12:10hrs, after finishing the telephone conversation with Zurich, the controller asked the crew, 'Are you sure you are over BN?' Capt Dorman replied, 'I think we have got a spurious indication, we are now on the LOC (short pause), on the ILS.' The controller replied, 'Ahh.'

At 09.12:33hrs the pilot confirmed, 'BN is established on glidepath and localiser; the ADFs (automatic direction finder bearings) are all over the place in this weather.' The controller replied that he could not see the Vanguard on his radar scope. At 09.13:03hrs Basel requested IM435's present height; both pilots simultaneously reported 1,400ft. The controller bit the bullet and said, 'I think you are not on the (short pause), you are on the south of the field.' This statement was not acknowledged, and all further calls to IM435 remained unanswered.

At 09.13:27hrs, the Vanguard, its undercarriage retracted and flaps set at 20°, brushed against a wooded ridge of hills in the Jura while apparently carrying out an overshoot and climb procedure. G-AXOP crashed and then somersaulted in the vicinity of Herrenmatt hamlet in the parish of Hochwald. 104 passengers and four crew died in consequence, but the absence of a major blaze, coupled with the fact that the rear fuselage section remained intact, ensured that there were 37 survivors. Of these, 35 passengers and one cabin attendant suffered injuries, but second stewardess Elisabeth Low escaped unscathed. Most of the victims were women from five Somerset towns who were on a one-day shopping expedition to Switzerland.

G-AXOP was a modest 11-year-old when it crashed. Examination of the wreckage showed no defects other than

Below: *A four-engined turboprop Vickers Vanguard, with three passenger compartments arranged for 139 tourist class passengers and four cabin crew, similar to that which crashed near Hochwald on 10 April 1973.*

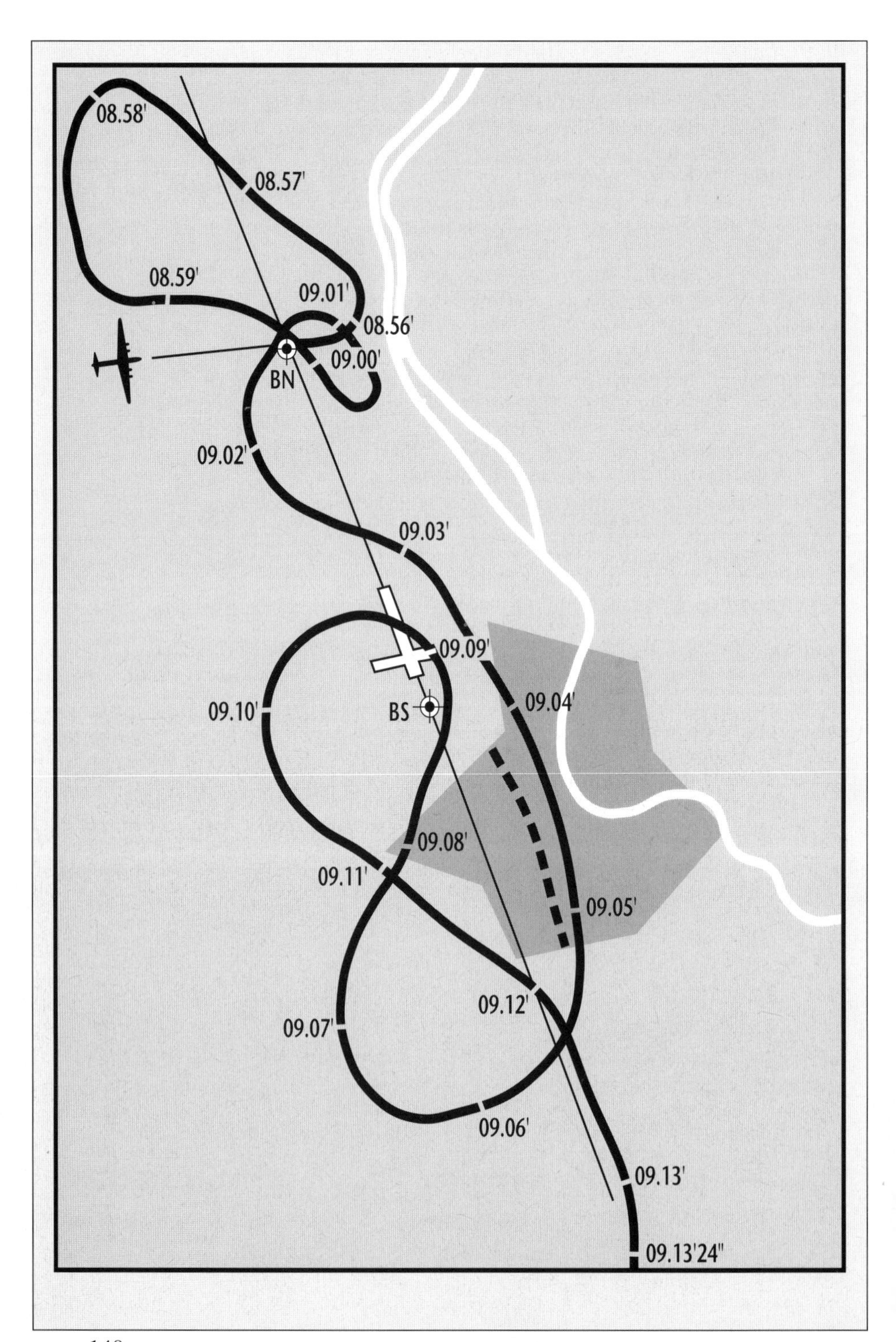
08.58'
08.57'
08.59'
09.01'
08.56'
09.00'
BN
09.02'
09.03'
09.09'
09.10'
BS
09.04'
09.08'
09.11'
09.05'
09.12'
09.07'
09.06'
09.13'
09.13'24"

those caused on impact, and the four Rolls-Royce Tyne engines were working symmetrically. The pilots were fit and rested, and neither was under the influence of alcohol or drugs. Yet G-AXOP ended up some 16km south of Basel airport, far away from where the crew thought it was.

Data from the FDR showed the transit from Bristol and descent to over BN to have been normal, but then the Vanguard's behaviour became 'highly erratic with regard to navigation and maintenance of height, deviating considerably from the prescribed flightpath and prescribed height'. IM435's meanderings resulted from the pilots becoming completely disorientated in bad weather. Although G-AXOP crashed in daylight, throughout its approach manoeuvres the Vanguard was almost continually in cloud. The cloudbase was low, visibility was greatly reduced in driving snow to approximately 150ft, and the wind was blowing from the north. At the airport, the RVR was below the minimum allowed for Vanguard operations, but the tower relayed an incorrect measurement (one slightly above the minimum) to the flight crew; they made no further inquiries regarding meteorological conditions.

Opposite: *The meandering flightpath of the Invicta International Airlines Vanguard. This illustrates the scale of the pilots' disorientation which ultimately led to the accident.*

In the accident area the cloud was continuous and visibility so poor that, when combined with snow-covered ground without contrast, the crew could not have seen whether they were approaching ground obstacles, particularly since the airliner was not fitted with a ground proximity warning device. Had the pilots been given accurate information on these conditions, they

Below: *The Hochwald crash site where 104 passengers and four crew died.*

may well have discontinued the approach to wait for a clearance, or made an approach to another airport where the weather was better.

The Swiss Federal Commission of Inquiry into Aircraft Accidents attributed the loss of G-AXOP to disorientation during two ILS approaches carried out under instrument flight conditions. A combination of unsatisfactory navigational procedures, confusion between navigation aids, and poor reception of radio-navigational-signals due to defects in the Vanguard's equipment added to the crew's problems. Air traffic control could also have done more to help and guide. However, the pilots had most to answer for. Capt Dorman had made 33 previous landings at Basel, while Capt Terry had made 61 landings there (14 of which were instrument approaches) including 38 on the Vanguard. Yet they screwed up in spades.

The crew's conduct following discontinuation of the first approach attempt remained inexplicable to the accident investigators. The crew knew that they had made a hash of the first approach, which had been flown by the 35-year-old Capt Dorman from the left-hand seat. Subsequent inquiries revealed that Capt Dorman's skills were not of the highest. He had been chopped from Royal Canadian Air Force flying training because of 'insufficient aptitude', his subsequent civilian flight training record was classed as 'dubious', his flying log book contained numerous 'discrepancies' and it had taken him nine attempts to secure his instrument rating. Capt Terry, on the other hand, was much more accomplished at the age of 47 and, after the first approach was thrown away, Capt Dorman's voice was heard on the radio which implied that Capt Terry had taken over the controls. The subsequent flight plot showed a marked improvement in the accuracy of flown turns and courses but still the second approach was 'bungled'.

The pilots apparently believed that their navigational difficulties were caused by atmospheric conditions — 'the ADFs are all over the place in this weather' — yet notwithstanding the prevailing unfavourable weather conditions and the instability of the flight instruments, the first approach attempt following confusion of the beacons need not have led to disaster. These conditions, some of which the crew were aware of, should have prompted them to make a careful, repeatedly verified approach with continual mutual monitoring during which any discrepancies would have been discovered and ATC help requested in case of doubt. But nothing of the kind occurred.

The two pilots held the rank of captain. In accordance with company practice, Capt Terry flew the first leg from Luton to Bristol, whereupon they swapped seats and Capt Dorman flew the second leg from Bristol to Basel. Perhaps the pilots' very experience told against them. A crew combination of this kind has been shown to have undesirable consequences, especially when it is not a training captain who is flying in the right-hand seat. The pilot in the right-hand seat then generally has insufficient experience of performing the co-pilot's real duties and thinks like a pilot-in-command rather than as a monitoring co-pilot. The aircraft commander in turn becomes irritated by the presence of a pilot of the same rank, so there is a danger of a crew made up in this way not acting as an effective flying team. If, after the first overshoot at Basel, Capt Terry took

Opposite: *A British Airways Boeing 747 similar to that which so nearly came to grief while trying to land at Heathrow on 21 November 1989.*

over control because Capt Dorman was making heavy weather of the whole business, Capt Dorman would have taken some time to alter his whole focus of attention, further degrading the checking and monitoring process. As proof of the pudding, some surviving passengers maintained that they briefly saw several houses on hilly ground during the second approach; the crew should have seen the same had they been alert, and mental warning bells should have rung given that the Vanguard should have been over flat, unbuilt-up land at the time. But the two minds on the flightdeck did not act in concert, and Flight IM435 came to grief in consequence.

The accident investigation report into the loss of G-AXOP recommended that all large commercially operated aircraft should be equipped with a ground proximity warning device. But technological wizardry counts for little if pilots do not have their wits about them.

On 21 November 1989, British Airways Flight BA012 undertook the long haul from Brisbane to London. A new flight and cabin crew had taken over the Boeing 747 (G-AWNO) in Bahrain after a rest of approximately 24 hours since their previous service, and the aircraft departed from Bahrain at 00.14hrs. After following a standard route, the crew received a terminal area forecast which indicated that CAT3 landing conditions (lower than 100ft cloudbase, 200m RVR) could be expected on arrival at Heathrow. Capt Stewart, in command, was assisted by a first officer who had not completed the additional training to enable him to operate in CAT2 or 3 conditions. Dispensation for the first officer to operate in all weather conditions was obtained from the 747 Flight Training Manager but in the event, as the 747 was in the hold at Lambourne, weather conditions improved to permit the crew to make a CAT2 (100ft cloudbase, 400m RVR) approach to Runway 27R.

G-AWNO left Lambourne with the 'A' autopilot engaged and it was vectored by Heathrow radar to intercept the ILS for 27R. 'Land' was selected some 13 miles from touchdown, the other autopilot was engaged and the 747 turned on to the final intercept heading of 240°. At this stage things were looking good. With its passengers strapped in, their trays stowed and their cigarettes extinguished, the great airliner descended to 3,000ft with flap 10 at about 180-190kt. It captured the altitude and then the localiser at 10 miles from touchdown.

The flight engineer saw a steady red warning on the 'B' autopilot, which he called. The FDR showed that shortly afterwards both autopilots were disconnected and the aircraft flown manually, but the crew had no subsequent recollection of this. They did remember that the 747 was not capturing very well, that it was going through the localiser and that the flight director was giving confusing information. The FDR showed a bank angle of between 15-17° during initial capture, that the aircraft was flown manually after autopilot disconnect with decreasing bank angle to just over two dots left of the localiser, and that the nav mode switch was selected to HDG and then back to either Land or ILS. This meant that the raw data display was showing displacement to the left of the centreline but that the flight director, which had lost its capture when the nav mode switch was cycled, was probably demanding a left turn when the crew would have been expecting a fly right demand.

Although he could not recollect doing so, Capt Stewart continued to fly the aircraft back towards the centreline. He selected 'A' autopilot just before G-AWNO crossed to the right side of the localiser at 2,470ft. At 1,670ft it turned back towards the centreline, recrossing it from right to left at 1,200ft when 'B' autopilot was again selected. During this gentle meander, the tower stated that half the supplementary high intensity narrow–gauge approach lighting was unserviceable. After referring to his *aide-mémoire*, the flight engineer told the pilots that they now needed a RVR of 600m.

After recrossing the centreline at 1,220ft, the aircraft continued out to the left before once again turning back slowly towards the centreline. Had the aircraft been flown manually on to the localiser and then aligned with it, and then the autopilots re-engaged, the 747 would have been expected to capture the ILS and achieve an autoland. But as G-AWNO did not meet the required parameters as 'B' autopilot was being reselected, an autoland was impossible. At 1,000ft, the crew should have abandoned the autoland and either gone round or continued to CAT1 minima (200ft/600m RVR). They continued their approach but did not adjust their minima. When the airliner completed its swing to the left, it recrossed the centreline from left to right at 577ft on a heading of 280° and continued to deviate at an angle of approximately 5°. At 500ft the ILS deviation lights illuminated, but the aircraft was allowed to continue to veer off to full scale deflection and to descend to 250ft for a full 17sec before the autopilots were disconnected.

Even after autopilot disconnection, G-AWNO was allowed to descend for another seven seconds until power was applied at 125ft and the aircraft began to pitch up. Nevertheless, it was rotated very slowly and therefore sunk to as low as 75ft in the transition process. By this stage the aircraft was passing over an access road to the A4, and staff and guests in the Heathrow Penta hotel could only watch in horror as the great 747 emerged from the clouds to miss them by an estimated 12ft. ATC was called at around 600ft and BA012 was cleared for a second approach to runway 27R. During repositioning, the weather improved to allow a CAT1 approach. Only the 'B' autopilot was used for this approach as the crew had doubts about the satisfactory performance of the 'A' autopilot. The aircraft, flown by Capt Stewart, landed without further incident.

This incident, which was only a whisker away from being a disastrous accident, was caused by a number of factors. The captain did not recognise that his aircraft had not become stabilised on the localiser. He did not discontinue the approach at 1,000ft despite failing to achieve the required conditions for an autoland. He did not respond immediately to the ILS deviation warning lights at 500ft with the aircraft diverging from the localiser centreline; on the contrary, he continued the approach to 250ft. He did not initiate an overshoot when it was clear that the 747 was so off-centre that a safe landing was out of the question. And after carrying on down for another 7-8sec, he initiated a go-around in a very tentative fashion such that a positive rate of climb was not achieved expeditiously. Inadequate monitoring during the approach, possibly caused by the first officer and flight engineer contracting gastro-enteritis some days earlier, and the distraction of an air traffic transmission, were contributing factors.

This was a particularly sorry incident, not least because Capt Stewart was subsequently found guilty of 'negligently endangering his aircraft and the 273 people aboard' at an unprecedented civil court hearing. After that ruling, it was not surprising that British Airways no longer felt inclined to retain Capt Stewart's professional services. Such was his subsequent sense of loss that Capt Stewart eventually committed suicide.

This human tragedy underlined the need for constant crew co-operation and monitoring on any aircraft flightdeck. The first requirement is that anyone assuming responsibility for aircraft and passenger safety must be physically and mentally fit for the task in hand. Then they must be trained properly, and the standard of that training must not only be validated and verified but also constantly refined and updated. But there are still examples of aircrew switching off to a breathtaking extent. On 5 September 1995, a McDonnell Douglas DC-10 belonging to Northwest Airlines crossed the Atlantic *en route* from Detroit to Frankfurt. The 241 passengers on board Northwest Flight 52 thought it strange as Brussels loomed into view as their destination on the cabin's electronic map. The flight attendants were also unsure what was happening — some even feared that a hijack was under way — but they all decided against talking to the pilots because of rules which forbade contact with the flight crew during an airfield approach. In the cockpit, the men at the controls only realised that they were landing at Brussels instead of Frankfurt when the DC-10 broke through the clouds. A source close to the subsequent incident investigation was quoted as saying, 'The only people on that plane who didn't know where they were, were the three guys up front.' It all came about because European air traffic controllers wrongly gave the crew directions for Brussels, and nobody on the flightdeck noticed.

But success or otherwise in a stressful flying endeavour such as landing a big aircraft with complicated systems in difficult conditions often comes down to human chemistry. At Cincinnati in 1965 and Basel in 1973, two aircraft crashed despite having two very experienced pilots up front; in effect, they failed to work as a co-ordinated team. Yet at Heathrow in 1989, a first

officer who was inexperienced on type may have been too physically rundown or too unsure to challenge the great man in the left-hand seat. Given that there is no guaranteed mix on any flightdeck, aircrew need to remain constantly aware that they are vulnerable no matter what their age and experience. Particularly when getting close to the ground, pilots should never take their eyes off the ball. An incident on 29 June 1994 proved that no amount of status on the flightdeck will offset a brain that is in neutral.

On that particular Wednesday morning in June, a BAe146 (ZE700) of the Queen's Flight was scheduled to fly HRH The Prince of Wales from Aberdeen to Islay off the west coast of Scotland. ZE700 left Aberdeen 23min late and was flown for most of the transit with the captain and co-pilot occupying the right- and left-hand seats respectively. The leg was too short to make up any significant time, and the crew resigned themselves to the fact that they would arrive late. About 10min before landing, whilst the captain was flying the descent, the Prince replaced the co-pilot in the left-hand seat. He flew the remainder of the approach and landing under the supervision of the captain, who was his personal pilot.

Islay, a small airfield consisting of Runway 13/31 and not much else, passed a surface wind of 250° at 20kt,

Above:
A BAe146 landing safely and wisely.

scattered cloud cover, and both runways — each offering a practical landing distance of 3,735ft — available for use. Because of cloud cover and high ground to the southeast, the captain elected to fly a procedural approach to Runway 13, retaining the option either to land straight-in with a tailwind or to circle and land into the wind on Runway 31. The captain briefed his intentions but made no mention of the tailwind to the Prince. To remain clear of cloud, HRH was directed to fly an abbreviated procedure which resulted in the four-engined 146 being lined up on final approach for Runway 13. It was then just under four miles from the threshold, above the normal approach path and fast; from this position, a steeper than normal final approach was required to reach the runway.

Twice during final approach, ATC passed the surface wind to the aircraft. The tailwind component of 12-13kt, combined with the steep approach, excess airspeed and a failure to settle the aircraft in the correct approach configuration, led to ZE700 crossing the runway threshold 30kt too fast. It touched down at 130kt, nose wheel first, with 2,244ft of runway remaining. The captain physically checked that HRH had selected the engines to ground idle and deployed the spoilers, and both pilots applied maximum braking. During the landing roll, one main wheel tyre burst and another partially deflated. Once the overheated brakes faded, there was nothing left to stop the 146 from overrunning. As the aircraft entered the runway extension, the captain tried to turn left on to the taxiway. However, the aircraft left the paved surface at slow speed, its nose wheel sinking into soft ground. ZE700's momentum slewed it around the nose oleo until it came to rest almost at right angles to the runway. The aircraft was shut down and the 11 occupants — six crew and five passengers — got out immediately. The Prince departed with measured calm for his programme of engagements.

Yet the implications of the Islay incident — headlines that could have screamed 'Heir to Throne Dies in Silly Accident,' not to mention the great expense of repairing ZE700 because aircraft are not stressed for pirouetting on their nose oleos — were such that a Board of Inquiry was convened. The Board found that the captain had made an error in his calculation of the maximum acceptable tailwind, and that the navigator had neglected to check the calculations independently. The captain had also planned to use Runway 13 because it offered an upslope, an easier NDB/DME approach for HRH to fly, and the lowest instrument approach minima. However, he failed to note that ZE700's landing weight — it had enough fuel on board to get back to Lyneham in Wiltshire — was well in excess of that for landing on 13, but within limits for 31. The Board concluded that the captain's decision to land on 13 was a major factor in the accident.

But the litany of failings only started there. A four-engined airliner weighing around 70,000lb is not a toy to be trifled with, yet far from sufficient time was allowed for the Prince to settle down and receive a comprehensive briefing after arriving on the flightdeck. The most critical factor, the tailwind, was not mentioned, and the final approach bore all the hallmarks of a classic 'rushed approach'. Standard 146 operating procedures for a premeditated 'steep approach' were not followed, even though they were even more essential given the tailwind; normal thrust was retained and only half airbrake used. At around 85ft, when the IAS was 141kt instead of the desired 109kt, thrust was reduced to flight idle. Thereafter, the rate of descent was reduced but still with the aircraft in a nose-down attitude.

Full airbrake was not selected over the threshold despite the excess speed, and the 146 decelerated slowly down the runway before landing nose wheel first. After several skips, the main wheels finally touched down with just over 1,500ft of landing distance remaining. At this point, the full range of retardation devices became available through the operation of the 'weight on wheels' logic. However, because the final skip had been of less than one second duration, the Dunlop brake system's design ensured that anti-skip protection was applied to the outboard main wheels only. As the pilots had already stamped on the brake pedals in the belief that they had landed, the inboard tyre on the right main gear burst at 102kt (with 750ft to run), and the left inner tyre deflated shortly afterwards. ZE700 eventually came to a halt with a collapsed nose gear after the nose wheels entered soft ground. The Board of Inquiry concluded that the mishandled approach and landing were major factors in the incident.

The Board determined that the captain's relationship with the Prince was one of instructor and student, and there was nothing in the personal relationship which would have prevented the captain from intervening or taking control. Throughout the approach, there was overwhelming evidence of handling errors that should have prompted a reasonable instructor to intervene and abandon the attempt to land. The cry of 'I have control' should have rung out loud and clear.

Yet despite having made a pig's ear of the approach, prompt intervention could still have stopped the aircraft in the landing distance available. Moreover, the captain elected to continue the landing despite reaching his predetermined 'go-around' point of 1,725ft without the 146 being properly on the ground. The inquiry concluded that the captain's failure to intervene and discontinue a grossly mishandled attempt to land caused the accident. He was found to have been grossly negligent. Despite being exonerated of any blame, to his credit the Prince of Wales 'grounded' himself from flying Queen's Flight aeroplanes ever again.

It was the captain's practice, when flying with HRH, to operate the 146 with the navigator rather than the co-pilot occupying the observer's seat. For his failure to check the captain's performance data, to advise him of the tailwind component, and to voice his concern during a patently inaccurate approach, the inquiry deemed the navigator to have been grossly negligent as well.

The captain and navigator had flown together for six years and this may have engendered a degree of mutual complacency wherein two very experienced aircrew felt reluctant, or did not consider it necessary, to challenge each other's actions or decisions. A good captain must always create a cockpit environment which encourages crew members to voice their concerns. Everyone in receipt of flying pay should monitor the pilot-in-command's flying and be ready to shout out if he or she looks like going awry. Whatever your aircrew specialisation on the flightdeck, you are never a passenger.

It is very easy to heap all the blame for any landing accident on to individuals, but it was not the wisest scheduling decision to send a BAe146 into a tuppenny-ha'penny airfield like Islay in the first place. Organisational support staffs must also remain alert to their responsibilities for flight safety. But in the final analysis, individuals such as Capt Stewart at Heathrow are made to pay the price for disastrous mistakes. So, if the crew bears ultimate responsibility, it must behave responsibly. It is easier said than done, but if the captain of the BAe146 going into Islay had been prepared to throw protocol-driven door opening times away and say 'Stuff the fact that HRH is late, flight safety is paramount,' it is almost certain that the accident would never have happened. Aircrew should never push, or allow themselves to be pushed, beyond their safe personal limits, whatever the deadline. Unless they mentally 'prepare for landing,' pilots may truly be embarking on their 'final approach'.

13 Afterthought

Flight safety has its lighter moments. In April 1995, a South African Airways Boeing 747-400 set off from London to Johannesburg with 300 passengers down the back and 75 breeding pigs in the hold. Two and a half hours out from London, the flight crew suddenly saw flashing lights indicating a fire hazard in the hold. The airliner was over Algeria at this stage but the captain decided to return to Heathrow. On landing, it was found that the severe stress of the flight had rendered the pigs so flatulent that their gas emissions had triggered the sensitive alarm system. Incredulous passengers, who were delayed for 15 hours, were only told of the cause of the alarm once the aircraft had resumed its flight to Johannesburg.

All of which shows that flight crews must remain on their guard for the unexpected. And they have good cause. New technology has solved many of the problems — such as failed engines and metal fatigue — that used to cause most crashes. But technology has also created new ways for things to go wrong. No fewer than nine of the latest generation of 'fly-by-wire' Airbuses, with 'glass cockpit' cathode-ray displays instead of traditional electro-mechanical dials, crashed between 1988 and mid-1995, raising awkward questions about the design of cockpits and the manner in which aircraft are flown jointly by human and computer.

Most crashes today are blamed on pilot error: an analysis by Boeing in 1993 showed flight crew behaviour to be the dominant cause of 60% of crashes. But this figure is somewhat misleading because it includes accidents which the crew failed to avert after something else had gone wrong first. The clinical answer would be to take humans out of the loop. Back in 1945, 244 RAF bombers sent to bomb Dresden were packed into an area just two miles wide and 1,000ft deep. That is barely half the space regarded as enough for just one airliner today, yet the bombers flew deep into European airspace without radar cover and through distracting anti-aircraft fire without wiping each other out. While there is no need to cram modern air traffic into such tight packets, we are in sight of the time when satellites and computers alone will be capable of controlling, navigating and separating the world's air traffic.

What will pilots do then? Good question. There will soon be no rational need for pilots in the air because aircraft will be controllable from elsewhere through data links. But the fare-paying public is unlikely to buy this arrangement, and the on-board calming presence and reassuring words of urbane and dashing Captain Courageous will continue to be *de rigueur*.

The great challenge for flight crews of the future will be to retain an interest in what is going on. As more and more airliners become programmed for automatic flight from take-off via a 8,000nm route to landing, pilots will find themselves increasingly responsible for computer

supervision rather than physical workload. Monitoring modern autopilots and other flight systems, which so rarely go wrong and which can often work more effectively than humans *when all is going well,* can be mind numbingly tedious especially at 02.00hrs after many hours' flying. Here lies the contradiction inherent in technology taking over from humans: the modern pilot's role is switching from performing to monitoring, but there is ample evidence, from both research and accident statistics, that humans are poor monitors.

And it does not take very long for some people to switch off. On 10 August 1995, a McDonnell Douglas MD-87 (SE-DHG) belonging to Scandinavian Airlines System took off from Gatwick on a positioning flight to Cardiff with two operating crew and 10 positioning crew on board. The weather at Cardiff was good, and the crew contacted Cardiff Approach Control at 11.36hrs as the aircraft passed FL60 in the descent to FL50. The controller instructed the crew to maintain FL50 on reaching it and to expect an ILS approach to Runway 12. After a further descent to FL40, the crew informed the controller that they had the airfield in sight and asked for a visual approach. The controller could not approve this request because of inbound VFR to the northwest, but he did clear the MD-87 to descend to 2,500ft saying that he would give a 'short radar to visual approach'.

Some 4nm to the west of the final approach to Cardiff lies St Athan, a military airfield housing the RAF's largest aircraft maintenance and repair depot. At 11.41hrs the Cardiff controller advised the crew that St Athan's 08/26 runway was active and that the MD-87 was to turn left on to a heading of 220°. The crew acknowledged this transmission. A few minutes later they were instructed to 'descend to 1,700ft, continue the left turn heading 150°, report field in sight'. The crew replied that they were still visual with the airfield. They were then instructed to contact Cardiff Tower;

Above:
An Eastern Airlines Lockheed L-1011 TriStar similar to that which crashed in the Florida Everglades.

the approach controller concluded his transmission with the words, 'Don't fly south of the final approach due to St Athan activity'.

At 11.43hrs, Cardiff Tower gave the crew clearance to land on Runway 12. A few seconds later the controller said to the crew, 'Confirm that you are not approaching RAF St Athan, you seem to be west of RAF St Athan at the moment.' The controller continued by transmitting, 'Break off and reposition on to Runway 12 at Cardiff. That is a military airfield you'll land up with.' The crew replied, 'Roger, we're breaking off'. They circled the airfield to the north and repositioned for an uneventful landing at Cardiff. Afterwards it was found that the approach charts supplied by SAS to the crew did not show the presence of RAF St Athan.

Little things like faulty charts can have a disproportionate effect on aircraft safety, but it is generally a chain of little things that build up into one big accident. Back in 1970, Eastern Airlines had signed up to be launch customer for the Lockheed L-1011 TriStar, a giant trijet offering a whole new standard of comfort, quietness and reliability. On 29 December 1972, Eastern Airlines TriStar N310EA found itself as Flight 401 on a non-stop domestic flight from wintry New York to Miami International. After an uneventful trip lasting less than 2.5hr, Capt Robert Loft told his delighted passengers, 'Welcome to Miami. The temperature is in the low 70s, and it's a beautiful night out there.'

At 23.32hrs, Miami Approach cleared Flight 401 to join for an ILS approach to Runway 09L. The TriStar was only 5nm to the northwest of the airport and in conditions of over 10nm visibility, Capt Loft was quite happy to call Miami Tower on 118.3MHz. While he changed frequencies, First Officer Albert Stockstill eased the great airliner leftwards to line up with the twin rows of lights marking 09L. Capt Loft's check-in transmission with the tower was blocked by a simultaneous call from another aircraft, so he turned to Stockstill and said, 'Go ahead and throw them out'. The first officer moved the undercarriage lever to the down position, and there followed the usual series of grinds and thumps as the three undercarriage legs went down. But only two instead of three 'greens' illuminated. 'No nose gear,' reported Stockstill.

Recycling the undercarriage had no effect, but there was no cause for alarm: the undercarriage was designed to be cranked down manually if all else failed. The problem was reported to the tower and the TriStar cleared to climb ahead to 2,000ft and return to Approach Control. Flight Engineer Don Repo then suggested that they test the indicator lights in case they were at fault. The undercarriage position indicators were mounted on the first officer's side of the flightdeck, and as Bert Stockstill found that he could not get a grip of the lights while flying manually, Repo moved forward to help. Leaning over Stockstill's shoulder, Repo pulled at the indicator assembly with equal lack of success. The commander should have taken over control of the airliner at this stage to free both of Stockstill's hands, but Bob Loft continued to talk to the approach controller.

Having overshot from Runway 09, Flight 401 was turned

Opposite: *The site of the first commercial wide-bodied jet fatal accident, looking back along the wreckage trail towards the point where Eastern 401 initially impacted. The biggest piece of wreckage — the tail section — is in the foreground. An idea of the scale can be gained from the man standing in the wreckage at right. The remains of the fuselage centre section and starboard wing are upper right.*

on to north. The time was 23.35hrs as Capt Loft became exasperated and told Stockstill to 'put the damn thing on autopilot' so that he could concentrate on the lamp assembly. N310EA was then turned on to 300° to keep separated from other traffic; after Stockstill complied, both pilots turned their attention to the errant light assembly. A minute later, Loft's patience ran out and he told Flight Engineer Repo to 'get down there and see if that damn nose wheel is down'. He meant that Repo should go down into the spacious electronics bay below the flightdeck, from where it was possible to check the gear's position through an optical sight. After further fruitless fiddling about with the peanut-sized light bulb, Loft ridiculed the fact that they were 'screwing around with a 20c piece of light equipment' on a wide-bodied airliner less than four months out of the factory which cost well over $10 million; everyone laughed.

As Miami Approach confirmed, 'We've got you headed westbound now, Eastern 401,' Don Repo reported from down below, 'I can't see it. It's pitch dark.' While the flight engineer continued to peer into the gloom, the approach controller noticed a height of only 900ft on Flight 410's alphanumeric block on his radar screen. He asked, 'Eastern 401, how are things coming along out there?' Unfortunately, he made no reference to the TriStar's altitude which continued to reduce.

Six and a half minutes had elapsed since the TriStar overshot and it was now well out over the totally dark and uninhabited Everglades to the west of the airport. 'Eastern 410 — turn left, heading 180', instructed the controller to swing N310EA back towards the airport. Stockstill coaxed the TriStar into another gentle turn but, as he did so, he sensed that something was not quite right. The ALT annunciator light in front of him was illuminated as it should have been, showing that the autopilot was still engaged in the 'Altitude Hold' mode, but as his gaze fell on the altimeter on the panel in front of him he blurted out, 'We did something to the altitude!'

'What?' queried Loft.

'We're still at 2,000 — right?' demanded Stockstill, unable to believe his eyes. There was a moment's silence on the CVR as Loft scanned his own instrument panel. Then he yelled, 'Hey, what's happening here?' Coincidentally the radio altimeter began to beep at an increasing rate. Just afterwards, the CVR picked up the sound of metallic impact out on the port wing. There followed a nightmare cacophony and then all was silence. The TriStar had crashed into the marshy Everglades some 20 miles west-northwest of Miami International. The time was 23.42hrs and what had started off as a seemingly minor annoyance ended up killing 103 people including the flight crew of three. A trail of fragmented wreckage extended for almost 500m in a southwesterly direction. Ironically, when rescue helicopters reached the area at first light, they reported that the airliner's three undercarriage legs had left deep tracks through the soft muddy swamp.

The shallow angle of the TriStar's impact with the swamp, the positioning of large cargo compartments beneath the passenger deck capable of absorbing much of the impact, and the robust nature of modern wide-bodied jet construction accounted for the almost miraculous survival of 73 people aboard Flight 401. But in among the wreckage, investigators found that the filaments of both nose gear light bulbs had simply burnt out. That being the reason for the start of the saga, the investigation then turned to the reasons for the airliner's unbidden descent from its holding altitude of 2,000ft.

NTSB investigators focused on three areas. The first, subtle incapacitation of the pilot flying the aircraft, was ruled out. Although post mortem examination revealed that Capt Loft had a tumour in his cranial cavity which could have affected his peripheral vision, it was concluded that this was not a factor in the events leading to the accident.

Secondly, there could have been a failure in the autopilot system. After the investigation it was found that the aircraft's two pitch control computers were mismatched. The control wheel force required to disengage Altitude Hold in computer 'A' on the captain's side was the correct 15lb, but for computer 'B' it was no less than 20lb. It was therefore possible, with the captain's autopilot engaged, to disengage 'A' computer but not 'B' computer. In these circumstances, the altitude mode light would have remained on while the ALT annunciator on the captain's panel went out. Misleadingly, however, the ALT annunciator would have remained illuminated on the first officer's panel, giving Bert Stockstill the false impression that the autopilot was still in the Altitude Hold mode. If Capt Loft had inadvertently pushed lightly on his control wheel as he turned round to speak to Flight Engineer Repo, he would have unknowingly disengaged the autopilot's Altitude Hold function, allowing the TriStar to lose height gradually.

Then came the penultimate link in the chain of 'gotchas' which turned an oversight into a tragedy. The CVR revealed that three minutes after the Altitude Hold disconnection, a half-second C-chord chime, indicating that the aircraft had deviated 250ft from the selected altitude, sounded on the flightdeck. But at this critical time, Flight Engineer Repo had descended into the electronics bay and both pilots were wearing headsets; consequently, none of the flight crew heard the warning chime.

This led on to the third and most convincing reason behind the airliner's unexpected descent into the Everglades — the crew had been distracted from monitoring the flight instruments adequately. Although the pilots lacked visual references by which they could have determined loss of height, it should have been evident that the TriStar was gradually descending by reference to either the captain's or first pilot's altimeters. But wider inquiries within the airline fraternity told NTSB investigators that generally, with the advent of new, advanced automatic flight control systems, crews were relying more and more on sophisticated computerised avionics, particularly as the reliability of new equipment increased. This meant that pilots were getting out of the habit of scanning basic instruments such as altimeters.

The flight engineer's attempts to see the nose undercarriage from down in the electronics bay must have distracted the pilots. Moreover, the crew were not as well briefed on the cockpit layout of the relatively new TriStar as they might have been. The flight engineer could not determine the true position of the nose gear because he could not see the leg. He could not see the leg in the darkness because the wheel well light had not been turned on. The crew believed that the wheel well light came on automatically whenever the undercarriage was down. This was not so; it was turned on manually, and the switch for doing this was on the captain's 'eyebrow' panel above the windscreen. If only someone aboard Flight 401 had known that.

There it was. After two years of safe operations by the new and much more reliable generation of jumbo jets, TriStar N310EA was overtaken by a disaster which did not result from any one major catastrophe or major error; rather it was the consequence of something as silly as a blown bulb compounded by several minor distractions from normal operating procedures. Such distractions on their own would have been of little account, but together they ensured that two highly qualified pilots — Capt Loft, in his mid-50s, had almost 30,000 flying hours while First Officer Stockstill, aged 39, was highly experienced with over 300 TriStar hours — devoted some four minutes to sorting out a bulb with minimal regard for monitoring their aircraft's flight situation. In the final analysis, neither pilot actually had responsibility for flying the TriStar over the Everglades.

So notwithstanding the superb technology incorporated into the magnificent Lockheed TriStar and its Rolls-Royce engines, despite its advanced, highly sophisticated, state-of-the-art automatic flight control system, Flight 401 fell victim to the most basic flight safety human error of all — the failure to maintain safe terrain clearance.

And maybe the crew knew this. Some of N310EA's black boxes and galley equipment recovered from the swamp was in such good condition that it was installed later in Eastern TriStar N318EA. During a flight on N318EA some months later, two flight attendants independently identified the face of Don Repo looking out from one of the ovens and warning of a future fire on the aircraft. Shortly afterwards, one of the TriStar's engines caught fire over Mexico.

On another occasion, N318EA was on a turnround in Newark for a departure to Miami when an extra passenger appeared in the first class cabin dressed in an Eastern Airlines captain's uniform. He sat there saying nothing to any of the flight attendants so the captain was called. He leaned forward to address his colleague, when he froze. 'My God,' he said, 'It's Bob Loft.' The mystery figure vanished.

Read into this what you will, but just as ghostly spectres transcend the ages, so basic flight safety truths remain constant. The advent of robust and reliable engines, resilient structures and uncluttered space-age glass cockpits must never seduce anyone into thinking that flying is becoming laughably easy. Whatever advances science and technology offer air travel in future, the ultimate responsibility of any pilot is to maintain control of his or her aircraft. This means knowing where you are, staying alert and resisting complacency at all times. Situational awareness is everything in the air, especially when things start to go wrong and the inevitable distractions crop up. If there is one abiding moral to take away from this book, it is this: there are no new accidents, only new pilots.

Select Bibliography

Birtles, Philip J.; *Lockheed TriStar*; Ian Allan Ltd, 1989
Gero, David; *Aviation Disasters*; Patrick Stephens, 1993
Jackson, A. J.; *British Civil Aircraft*, vols I-III; Putnam, 1988
Job, Macarthur; *Air Disaster*, vol 1; Aerospace Publications, 1994
Janes' All The World's Aircraft; Janes Publishing Co, various editions
Launay, Andre; *Historical Air Disasters*; Ian Allan Ltd, 1967
Sereny, Gitta; *Albert Speer*; Macmillan, 1995
Wright, Alan J.; *Classic Civil Aircraft: Boeing 707*; Ian Allan Ltd, 1990
Wright, Wilbur; *The Glenn Miller Burial File*; 1993
After The Battle magazine, Pt 39

Finally, *Flights To Disaster* could not have been written without access to the detailed investigations and conclusions contained within AIB, AAIB, CAB and NTSB accident reports. I am eternally grateful to the men and women who painstakingly compiled them, and if any false conclusions are drawn from their labours, the fault is mine alone.

Index